Kentucky Nature Studies : 2

Amphibians & Reptiles of Kentucky

AMPHIBIANS & REPTILES OF KENTUCKY

Roger W. Barbour

The University Press of Kentucky

Standard Book Number 8131-1233-8

Library of Congress Catalog Card Number 71-132823

A statewide cooperative scholarly publishing agency
serving Berea College, Centre College of Kentucky,
Eastern Kentucky University, Kentucky State College,
Morehead State University, Murray State University,
University of Kentucky, University of Louisville,
and Western Kentucky University

Editorial and Sales Offices: Lexington, Kentucky 40506

For Marsha, Bill, & Jim

Contents

PREFACE

The amphibians and reptiles of Kentucky are poorly known, and no comprehensive treatment of them exists. This volume represents an attempt to fill that need in a manner that will be useful not only to the student but also to anyone interested in natural history. It is my intention to provide sufficient material in usable form to enable such persons to acquire an understanding of our native amphibian and reptilian fauna. To that end I have tried to present here for each species occurring within Kentucky the pertinent information on its recognition, variation, confusing species, habitat, Kentucky distribution, and life history. Identification of species and subspecies is facilitated by illustrated keys as well as by photographs in color. Insofar as possible technical language has been kept to a minimum.

The need for this volume is evident in the increasing number of requests for information about our amphibians and reptiles. Heretofore little of this information was readily available because the knowledge was either unpublished or widely scattered in the literature, some of which is out of print. The preparation of this book was also prompted by my long-held belief that any real progress in the conservation of our natural resources must come from the grass roots of the nation and that an awareness of the necessity for conservation will only come through knowledge of our environment. Of this environment amphibians and reptiles are an inconspicuous but important part.

The color photographs save two illustrating this volume are by the author. I wish to thank Tom C. Fuller for the photograph of amphiuma, and Dr. Robert S. Simmons for his photograph of the coal skink.

Especial thanks are also due the Kentucky Department of Fish and Wildlife Resources for permission to use their color separation negatives of some of my photographs; the Illinois Natural History Survey for permission to use some of their line drawings; Mike Creusere for information on *Ambystoma platineum;* Faith Hershey for preparing the pen and ink sketches; and Barbara Woodruff Birge and Julia Wade for typing the manuscript.

I should like to acknowledge my indebtedness to the many students and friends who, over the years, have contributed to this volume. To the Research Committee of the University of Kentucky I am grateful for several grants over the past twenty years that have in part made possible the research on which this publication is based.

INTRODUCTION

The prominence of amphibians and reptiles in the vertebrate fauna of Kentucky belies somewhat their minor position on a world-wide basis. The state has a relatively rich herpetofauna; 99 species occur, 45 amphibian and 54 reptilian. Counting subspecies, the total number of forms of amphibians and reptiles known to occur within the state totals 122. Although none of the recent species are positively known to have been extirpated from the state, the status of one is unclear, and the position of several is precarious.

Several factors contribute to the large number of reptiles and amphibians in Kentucky, of which geography is perhaps the most important. Some northern, southern, eastern, and western species range into the state by virtue of its central location. The western tip of the state, extending into the Mississippi embayment, includes the northern portions of the range of a number of species typical of the southern coastal plain. The eastern end of Kentucky, extending into the Cumberland Mountains, includes the more southerly portions of the range of some northern species.

The great diversity of habitats, ranging from low-lying cypress swamps characteristic of the Deep South through a great variety of aquatic, herbaceous, and wooded areas to the moist beech-birch-maple climax forests of the higher slopes of the Cumberland Mountains, is a second major factor contributing to the abundance of Kentucky's herpetofauna.

Although evidences of man's activities are abundant in nearly every available habitat, many areas suitable for the existence of reptiles and amphibians still exist. Unfortunately, they are being destroyed at an ever-increasing rate, and the future of many of these small, innocuous animals is uncertain.

While it is true that many people have never seen a salamander (although they are quite common), the voices of the frogs and toads on warm spring nights constitute a very large segment of the audible out-of-doors. Some of the frogs are classed as game animals and offer considerable sport and delicious eating. Several of the turtles are also excellent to eat, and even more are kept as pets. Except by country boys and zoologists, lizards are largely ignored. Snakes, on the other hand, command a lively interest. Fortunately, there seems to be a definite trend toward a more wholesome, less morbid, view of these fascinating animals.

Amphibians and reptiles are of minor economic importance. We eat some, we keep some as pets, some of them eat noxious insects and other pests, but on the whole they have little direct effect on Kentuckians. A few of the snakes are venomous, but they are scarce, generally inhabit the more remote areas, and usually flee from man. The harm they do is negligible, far out of proportion to their fearsome reputation.

Neither amphibians nor reptiles generally have much esthetic appeal; in this respect they are far outclassed by both birds and mammals and even by the more colorful fishes. This is due in part to ignorance: most people simply are not aware of the great diversity, the attractive coloration, or the interesting behavior of these animals. Many amphibians and reptiles are uncommonly handsome, rivalling the birds and far surpassing the mammals in the variety of their colors and markings.

In their own right, and quite aside from any influence, good or bad, that they might have on mankind, the amphibians and reptiles are worthy of our study, consideration, and protection. They are among the most ancient of vertebrates inhabiting the earth today—only the fishes and their relatives are older—and exhibit in more primitive form many of the characters and characteristics of the higher vertebrates. From a behavioral standpoint, they go their way, living out their lives for the most part quietly and inconspicuously. They are constantly influenced by their environment; every living thing with which they come in contact has some influence, however subtle, upon them. The physical features of their environment exert a tremendous influence upon them; in many cases even when and where they breed is determined by the physical factors of their environment. In turn, they influence all life about them, and to a lesser extent, even the physical features of the area in which they live. In essence, the amphibians and reptiles are inextricably bound up in the extremely complex and sometimes fragile web of life on this planet. So interlocking is the web, and so all-inclusive, that figuratively, at least, you jar the entire planet when you pull the tail of the littlest salamander.

HOW ANIMALS ARE CLASSIFIED

By way of an introduction to what follows on the use of this book and the keys, the scheme of classification of animals is reviewed briefly below. The science of classifying organisms is *taxonomy*.

1. All animals, of whatever kind, are grouped together in the kingdom ANIMALIA.

2. The kingdom is divided into major groupings, called PHYLA (singular—phylum). There are about 40 phyla, but most of the commonly encountered animals are members of about 10 phyla. For example, all animals possessing gill slits at some time and a vertebral column, or notochord, are members of the phylum CHORDATA. This phylum includes all the fishes, amphibians, reptiles, birds, and mammals, as well as a few more primitive forms.

3. Phyla are divided into classes. The bony fishes are in the class TELEOSTOMI; the amphibians, in the class AMPHIBIA; reptiles, in REPTILIA, etc.

4. Classes are divided into orders. The class AMPHIBIA, for example, is divided into the order ANURA (frogs and toads), the order CAUDATA (salamanders), and others.

5. Orders are divided into families. The order ANURA is composed of the families HYLIDAE (treefrogs), BUFONIDAE (true toads), RANIDAE (true frogs), etc.

6. Families are divided into genera (singular—genus). For example, the family HYLIDAE is represented in Kentucky by 3 genera—*Hyla* (the treefrogs), *Acris* (the cricket frogs), and *Pseudacris* (the chorus frogs).

7. Genera are divided into species (species—both singular and plural). There are several species of *Hyla* in Kentucky: *Hyla versicolor* (the gray treefrog); *Hyla crucifer* (the spring peeper); *Hyla cinerea* (the green treefrog); etc. The species concept is difficult to define. Essentially, a species is a specific kind of an organism. Members of a species readily breed together but not with other species; if they do crossbreed, the hybrids usually are sterile.

The scientific name of an animal species consists correctly of 3 parts, for example *Hyla crucifer* Wied. The first part, always capitalized, is the genus. The second part, never capitalized, is the trivial name. The third part, often omitted, is the name of the person who first recognized and gave the animal the trivial name. Sometimes the author's name is in parentheses; in this case, it simply indicates that when he initially named the species, the generic name was not the same as it is now.

Any of the groups of animals (taxa; singular—taxon) can be divided into subgroups, as subphyla, subclasses, etc. These groups pose few problems and offer no confusion until we arrive at the species level.

A subspecies is simply a part of the total population of a species—a population that is separable from other populations, that occupies a definable geographic area, and that intergrades in a broad or narrow belt with other populations of the species where they meet. Obviously 2 subspecies of a single species cannot occupy the same geographic area, unless they are separated by some habitat, behavioral, or temporal difference. It is rare indeed to encounter such situations.

The arrangement of this book follows the taxonomic scheme outlined above, and the various taxa are organized according to a currently accepted scheme of evolutionary sequence; more primitive animals appear first, followed by those of increasing advancement.

The function of this volume is twofold: to enable a nonprofessional to identify any amphibian or reptile native to Kentucky and to enable him to learn something of its distribution and life history. Obviously, identification of the animal in hand is the first and most difficult step; it may be accomplished in two ways.

By direct comparison with the color photographs, one can arrive at the proper identification in most cases, near it in all. By then reading the sections on recognition, variation, and confusing species, one can arrive at the correct identification. Partial verification can be obtained by reading the sections on habitat and distribution and by checking the range map.

The second method of identification involves the use of the illustrated keys; although slower, it is the more accurate and informative of the two systems. Verification of the identification can be obtained by comparing your specimen with the photographs and with the written description.

The keys are nothing more or less than a series of numbered couplets, each offering a choice of characteristics headed *a* and *b*. By choosing the correct description, time after time, one is directed forward by number through the series of couplets until he arrives at the name of the animal in question. The small photographs with the key will be of value in making the correct decision.

The first choice to be made is whether the animal is an amphibian or a reptile. Necessary information to make this choice is presented on page 6. There, one is directed to the proper class (Amphibia or Reptilia).

Following the discussion of the class is a key to the orders of the class, and when the proper order is chosen, one is directed to the order.

Following a description of the order is a key to the families represented in Kentucky. Here again, by making the proper choices, one is directed forward, to the treatment of the family. With a little experience and familiarity with the animals and the scheme of classification, one can turn directly to the proper family without bothering with the keys.

When the correct family is determined, the key presented there will enable one to identify an animal to species or to subspecies if more than one subspecies occurs in the state. When the name of the animal is finally reached, one is directed to the page where the discussion of that animal occurs.

In the species accounts, especially in the sections on distribution in Kentucky, frequent mention is made of county names and physiographic provinces. These are indicated in Figs. 1 and 2. The Roman numerals after the common name indicate the appropriate color plate.

HOW TO DISTINGUISH AMPHIBIANS AND REPTILES

Nearly every vertebrate animal inhabiting Kentucky is a member of one of only 5 classes (the lampreys, in a class of their own, are the exception). Fishes, amphibians, reptiles, birds, and mammals are the animals; Teleostomi, Amphibia, Reptilia, Aves, and Mammalia are the corresponding classes.

There are many ways in which the classes differ, but most of them can be quickly recognized by a single characteristic or so. Mammals have hair, birds have feathers, fish have fins; these classes are easy, offer no problem, and should never be confused with amphibians or reptiles.

The amphibians occurring in Kentucky all have a completely scaleless skin, which is most often moist and slimy, but sometimes warty and quite dry. They all have at least 2 legs, and the toes are never armed with nails or claws. The body may be short, stout, and tailless with strong hind legs (frogs and toads), elongate, tailed, and slender with 4 legs (salamanders), or elongate with only 2 legs (trachystomes). The class Amphibia is discussed starting on page 13.

Reptiles all have scaly (leathery, in the softshell turtles) skins, which is not at all slimy. Some have legs, some not, but when legs are present the toes are always armed with claws. The body may be rounded (turtles) or elongate (snakes and

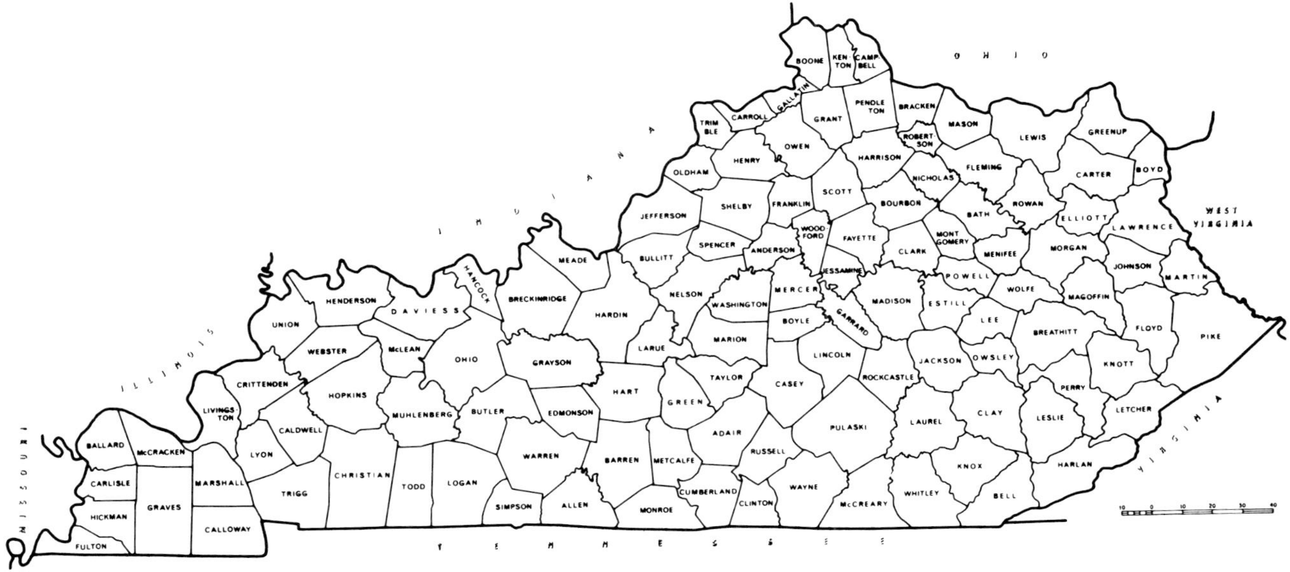

Figure 1. KENTUCKY COUNTIES

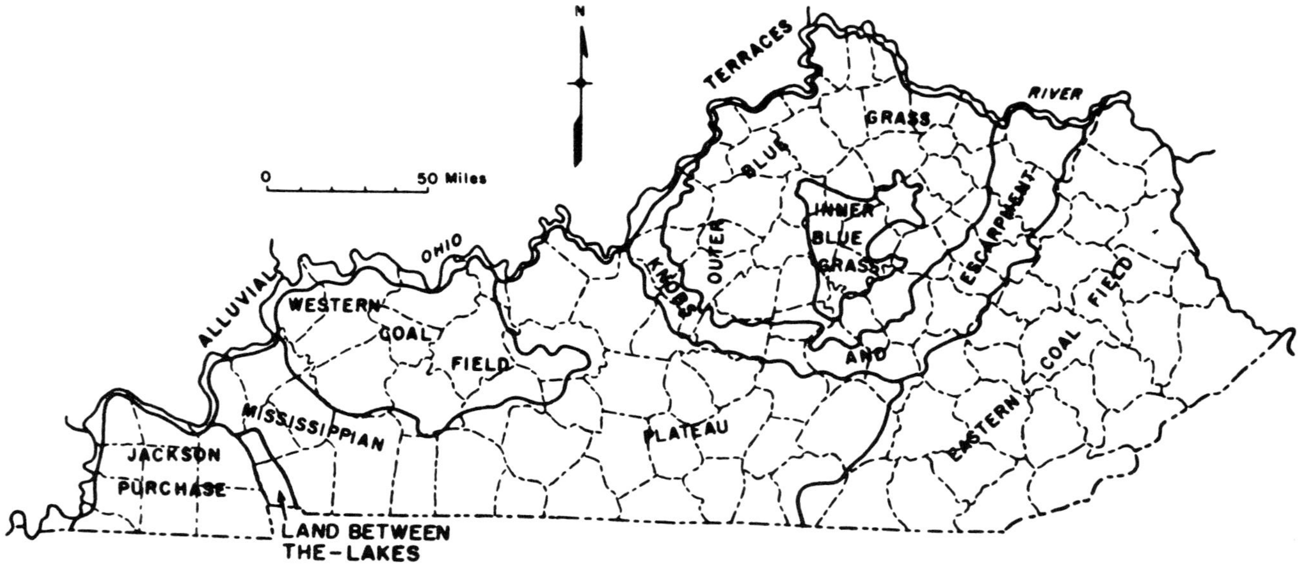

Figure 2. PHYSIOGRAPHIC PROVINCES OF KENTUCKY

lizards). Lizards have eyelids and external ears and usually have legs (one occurring in Kentucky is legless); snakes have no eyelids, no external ears, and no legs. The class Reptilia is discussed starting on page 145.

There is really only one problem in identifying the major groups of our amphibians and reptiles: salamanders are confused with lizards. To reiterate: Salamanders have scaleless, moist skins, never have claws on the toes, and the vent is a longitudinal slit; Lizards have scaly, dry skins, and their toes, if present, always have claws, and the vent is a transverse slit.

Amphibians

AMPHIBIAN TOPOGRAPHY

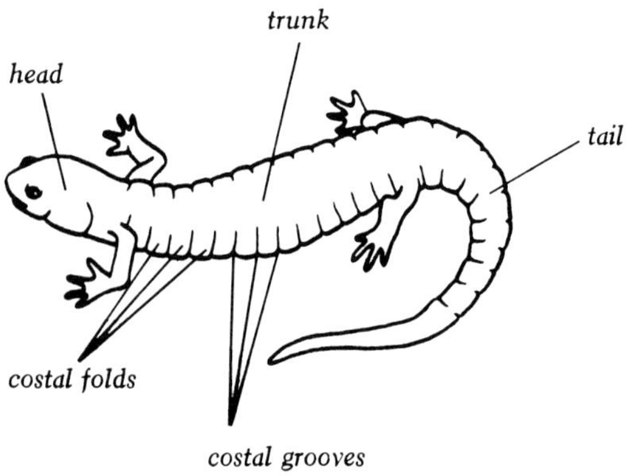

SALAMANDER

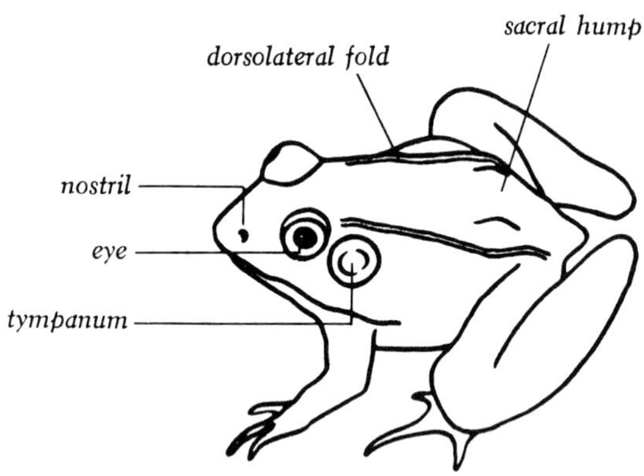

FROG

AMPHIBIA

Unfortunately, there is no single characteristic, such as the hair of mammals or the feathers of birds, which we can point to and say, "All amphibians have this; no other animal does."

Members of the class Amphibia are generally small, innocuous animals with smooth, moist, scaleless skins. Most have feet with toes, but in no case do the toes have claws. Their body temperature fluctuates with that of the environment; hence they cannot long survive below freezing temperatures and are less active at low temperatures.

Amphibians are surprisingly variable. The name, *Amphibia*, means "double life," in allusion to the fact that, typically, part of the life is spent in water and part on land. It is quite true that many amphibians lay their eggs in water, and the hatchlings go through a larval period as gill breathing, aquatic animals. They metamorphose, after a period of time from a few days to a few years, to become air-breathing, terrestrial adults. This "typical" life history has many variants, however. Some species retain the eggs within the body of the female until they hatch, and the youngsters are then born as miniature editions of the adult. Some lay eggs on land, and the larval period is spent in the egg; the hatchlings are like tiny adults. Many other variations occur.

Although not all amphibians lay their eggs in water, the eggs must remain in a moist place if they are to hatch. They do not possess a shell, but are surrounded by coats of a jelly-like substance. These coats, although somewhat protective in nature, are of little value against desiccation.

Amphibians are now a minor group of vertebrates, limited to about 3,000 species. Such was not always the case. They were the first vertebrates to successfully colonize the land, and became the dominant animals upon the face of the earth some 300 million years ago, and occupied that position for nearly

100 million years. During that time at least 9 orders existed which are now extinct. The fossil record indicates that the oldest currently existing order, the Anura, arose some 180 million years ago.

Four orders of Amphibia exist today. One of these, the Apoda, is composed of long, slender, wormlike, blind, burrowing creatures without limbs or limb-girdles, called caecilians. These rarely reach a length of over a foot. A few primitive genera have minute scales imbedded in the skin. There are about 75 species known, occurring in tropical areas, usually in forests or along forest edges.

The other three orders have living representatives in Kentucky. Members of the order Trachystomata, the trachystomes, are elongate, aquatic creatures with minute front legs and no hind legs at all. The Caudata are the salamanders, with 4 legs, elongate bodies, and an obvious tail. The Anura, the frogs and toads, are short, squat, 4-legged, and tailless.

KEY TO THE ORDERS OF KENTUCKY AMPHIBIA

1 a. Body short and squat, tailless / Order ANURA, Frogs and Toads (p. 91)

b. Body elongate with a tail 2

2 a. With only an anterior pair of limbs / Order TRACHYSTOMATA, Sirens (p. 16)

b. With 2 pairs of limbs (may be minute) / Order CAUDATA, Salamanders (p. 18)

1a

2a

2b

ORDER TRACHYSTOMATA Trachystomes

This is the smallest order of the amphibians, with but one small family, found only in the central and southeastern United States and northeastern Mexico.

The trachystomes are unlike any other amphibian. All are aquatic, permanent larvae, with tiny, lidless eyes. The front limbs are so small as to be nearly useless, and the hind limbs are lacking. Although the animal has well developed external gills, it also has lungs.

FAMILY SIRENIDAE Sirens

This small New World family is made up of only 3 species in 2 genera, one of which occurs in Kentucky. All members of the family are permanently aquatic and have bushy red external gills. The legs are much reduced, and only the front pair are present.

Siren 1

Siren intermedia Le Conte

Recognition: An elongate, eellike animal, with a single pair of limbs. Permanently aquatic, with 3 pairs of external, bushy red gills. Length usually 10 to 15 inches, up to 16⅞. Color various, gray, brown, olive, or black; sometimes with scattered spots or flecks. Belly lighter than dorsum.

Variation: No geographic variation has been observed in Kentucky; *Siren intermedia nettingi* Goin, the western lesser siren, is the only subspecies that occurs in the state. Males are larger than females, but it is difficult to determine the sex without dissection.

Confusing Species: Most like Amphiuma, which has 4 limbs, albeit diminutive ones, and no gills. Sometimes confused with the eel, a fish, but this latter is markedly different with its scales, internal gills, and fins.

Habitat: Muddy swamps, cutoffs, sloughs, drainage ditches, and similar sluggish waters. Cypress swamps seem favored in Kentucky.

Kentucky Distribution: Poorly known, because of its secretive nature. Common throughout the Jackson Purchase but the eastward distribution in the state is unknown. It is known from Hopkins County and probably occurs in suitable habitats at least as far east as Butler County. The species is abundant in some areas; in Hickman County, it makes up about a third of the total food of the cottonmouth moccasin.

Life History: Although this species remains active most of the winter, it spends at least some time in dormancy. In early

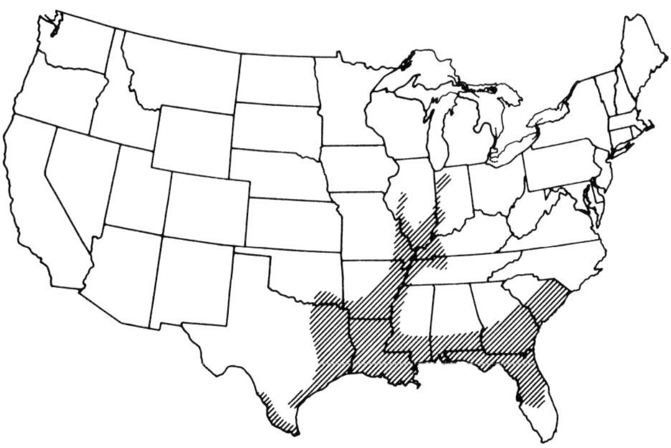

Range of the Siren in the United States

spring the females deposit several hundred eggs in hollows in the muddy bottoms of ponds or streams. The individual eggs are about 1/8 inch in diameter. Little information is available on incubation period or larval development, but one youngster was known to have increased its length nearly fourfold in 6 weeks.

Food consists of a variety of items, but apparently is chiefly arthropods and worms. Occasionally a bank fisherman in western Kentucky takes a siren on a baited hook. When handled, sirens wriggle violently and are so slippery that they are almost impossible to hold without seriously injuring them.

Remarks: So little is known of this animal that any accurate observations on its life history would be of value; a serious study of the life history would be rewarding.

ORDER CAUDATA Salamanders

Unlike the frogs and toads, salamanders retain their tails throughout life, hence the name. Some authors prefer the term *Urodela* for the ordinal name; the meaning is the same: Caudata = tailed, Urodela = with a visible tail.

Salamanders have an elongate body, 2 pairs of relatively weak legs, and a more or less distinct neck. Most are oviparous, and fertilization is usually internal, by means of spermatophores.

Although tied to a moist environment, they have exploited it well. Some are permanently aquatic; some have both aquatic and terrestrial phases in the life cycle; and some are completely terrestrial. A few are partially arboreal, and some have forsaken the surface to live permanently in underground waters.

Salamanders exhibit various means of obtaining oxygen. Some have permanent gills, some have both gills and lungs, some have first gills and then lungs, some gills and then neither gills nor lungs. These latter, the Plethodontidae, are quite successful, and by far the most abundant family.

Approximately 280 species of Caudata are known, in 7 families. Six of these families occur in Kentucky, represented by 27 species in 12 genera.

KEY TO THE FAMILIES OF KENTUCKY CAUDATA

1 a. With a wrinkled, fleshy fold of skin on each side of trunk; head broad and flat, mouth wide; tail flattened from side to side; permanently aquatic / Family CRYPTOBRANCHIDAE, Hellbenders (p. 21)

1a

b. Without a wrinkled, fleshy fold of skin on sides of trunk; head not particularly flattened; tail often rounded, but sometimes keeled or flattened; aquatic or terrestrial 2

2 a. Limbs minute, apparently functionless; body extremely elongate, eellike / Family AMPHIUMIDAE, Amphiumas (p. 43)

b. Limbs obviously functional; body elongate, but not eellike 3

2a

2b

3 a. With a tiny, but well-defined groove from each nostril down to the lip / Family PLETHODONTIDAE, Lungless Salamanders (p. 45)

b. No groove from nostril to lip 4

3a

4a

4 a. Four toes on each hind foot; adults permanently aquatic, with bushy external gills / Family PROTEIDAE, Mudpuppies (p. 88)

b. Five toes on each hind foot; adults without external gills; adults may occur in the water, but are not permanently aquatic 5

5 a. Costal grooves present / Family AMBYSTOMATIDAE, Lunged Salamanders (p. 24)

b. Costal grooves absent / Family SALAMANDRIDAE, Newts (p. 40)

5a

5b

FAMILY CRYPTOBRANCHIDAE Giant Salamanders

This small family embraces not only our most primitive salamander, but also the largest existing amphibians. There are but 3 species in 2 genera in the family; one genus (*Megalobatrachus*) with 2 species occurs in Japan and China; the other genus, *Cryptobranchus*, with but a single species, is limited to eastern United States.

Although our species, *C. alleganiensis*, rarely attains a length of over 27 inches, the oriental species sometimes attain a length of 5 feet or so.

These salamanders are permanently aquatic and evidently undergo only a partial metamorphosis. The adults lose the gills at metamorphosis, and though the gill slits on both sides close in *Megalobatrachus*, usually in *Cryptobranchus* the slit closes on only one side. The adults never develop eyelids, but retain the lidless characteristic of larval salamanders, as well as the larval dentition.

This is the only family of our salamanders in which fertilization is external.

Hellbender I

Cryptobranchus alleganiensis (Daudin)

Recognition: A huge, flattened, grotesque, aquatic salamander. No external gills, but having a thick, fleshy fold of skin along each side and on the top and bottom of the tail. Head extremely flat; mouth wide with many small teeth. Eyes minute, inconspicuous. Length: male, usually 10 to 18 inches, to 22; female, usually 12 to 24 inches, to 29⅛. Color dark, yellowish brown to almost black, uniform or with small scattered irregular flecks and spots, sometimes light, sometimes dark.

Variation: The subspecies *Crytobranchus alleganiensis alleganiensis* (Daudin) is the only race recognized in Kentucky. However, individuals from western Kentucky often have more abundant, larger and darker spots than those from more eastern sections of the state. Males are smaller than females and have a swollen ridge around the vent in the breeding season.

Confusing Species: Sometimes confused with the mudpuppy, but the latter has bushy external gills.

Habitat: Larger streams and rivers where large stones provide daytime retreats and nesting sites. Sometimes found in small, clear, rocky streams.

Kentucky Distribution: Statewide except western Kentucky west of the Cumberland River. It seems most common in the upper reaches of the Cumberland, Kentucky, and Licking river systems.

Life History: In late August or September males take up residence in saucer-shaped nests excavated in the stream bottom beneath a large flat rock or other sheltering object. A female may lay as many as 500 eggs in the nest, and several females may use the same nest. Fertilization is external, unlike any other native American salamander. The male often lies among the eggs with his head guarding the entrance to the nest cavity.

The pale, yellowish eggs are laid in long strings and form a tangled mass at the bottom of the nest. The individual egg is about ¼ inch in diameter, but its surrounding membranes and fluids bring the total diameter up to about ¾ inch.

Incubation is from 2 to 3 months, and the eggs usually hatch in October or November. At hatching the youngsters are nearly 1½ inches long, with external gills. They lose the gills when about 5 inches long and approximately 18 months old.

A bottom feeder, the hellbender will eat almost any animal matter, whether fresh or putrid, and it is frequently caught on a hook. Many fishermen consider it to be exceedingly poisonous and would not voluntarily touch one under any circumstances.

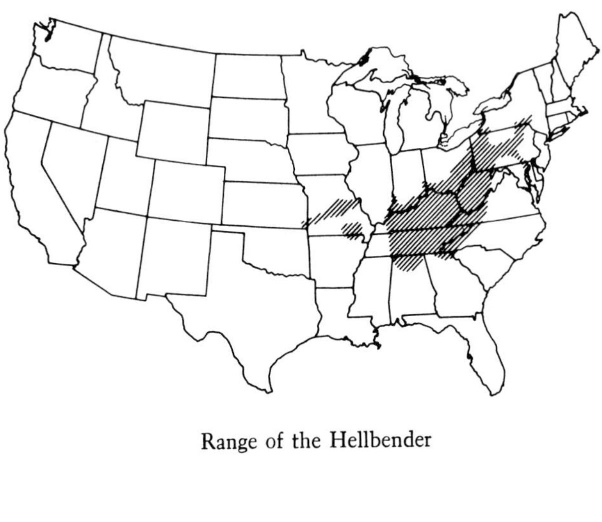

Range of the Hellbender

Actually, the hellbender is not poisonous at all and is not particularly vicious, although it will bite savagely when sufficiently provoked. If grasped firmly behind the head, imprisoning the fore limbs, it is completely unable to turn and bite and may be handled with impunity.

There is little doubt that the hellbender competes with some fish, particularly catfish, for food. However they rarely, if ever, attain a population sufficiently high to offer any serious competition for food. In fact, the presence of hellbenders in a lake or stream may be of actual benefit, for they are frequently eaten by large catfish.

FAMILY AMBYSTOMATIDAE

This family is composed of 6 living genera, 3 of which occur in the United States. It is represented in Kentucky by one genus, *Ambystoma*, and 7 species. Fertilization is internal, and the eggs are usually laid in water. The larvae are gilled and go through an aquatic phase before metamorphosing into the adult stage. Adults have well-developed lungs. In some areas, some species are known to breed in the larval stage (neoteny), but this phenomenon has not been recorded for any species of this genus in Kentucky.

KEY TO THE SPECIES OF *Ambystoma*

1 a. With yellowish to orange spots or blotches on body 2
b. Variable, but without such spots 3

2 a. With rounded spots arranged in rows down back; belly gray / *Ambystoma maculatum* (Shaw), Spotted Salamander (p. 36)
b. With irregular spots, not arranged in rows; belly greenish-yellow with dark mottlings / *Ambystoma tigrinum tigrinum* (Green), Eastern Tiger Salamander (p. 38)

2a

2b

3 a. Dorsal pattern of bold silvery or gray cross-bands / *Ambystoma opacum* (Gravenhorst), Marbled Salamander (p. 34)

b. Dorsum unicolored, or at least without silvery cross-bands 4

3a

4a

4 a. Costal grooves 10 or 11, counting one each in the axilla and the groin; body short, dumpy; tail short; head large / *Ambystoma talpoideum* (Holbrook), Mole Salamander (p. 33)

b. Costal grooves 12 or more; body, tail, and head more reasonably proportioned 5

5 a. With grayish lichenlike patches scattered over sides; head small, rounded / *Ambystoma texanum* (Matthes), Small-mouthed Salamander (p. 31)

b. With blue or blue-white flecks or spots scattered over sides; head larger, flatter (two species of the Jefferson salamander complex key out here; they are impossible to separate reliably without recourse to a microscope) 6

6 a. Chromosomes 28; erythrocytes 600-860 square micra; gray to gray-brown, nearly unicolored with extremely long toes or with many fine bluish flecks on sides / *Ambystoma jeffersonianum* (Green), Jefferson's Salamander (p. 26)

b. Chromosomes 42; erythrocytes 920-1510 square micra; gray to gray-brown, with many larger flecks and spots on sides; unisexual, females only / *Ambystoma platineum* Cope, Triploid Jefferson's Salamander (p. 29)

Jefferson's Salamander II

Ambystoma jeffersonianum (Green)

Recognition: A relatively slender, dark salamander, often with extremely long toes. Length usually 5 to 7 inches, to 8¼. Color brown or gray, belly lighter. Often with small bluish flecks, particularly on the sides of the head, body, and tail. Costal grooves usually 12.

Variation: No geographic variation is recognized. The males can be recognized in the breeding season by the swollen vent, but at other seasons the sexes are externally similar.

Confusing Species: Most often confused with the triploid Jefferson's salamander (see following account). Small-mouthed salamanders have much shorter toes, and shorter, broader, more rounded heads. Nearly unspotted slimy salamanders are somewhat similar but have a groove from the lip to the nostril.

5a

6a

6b

Habitat: Woodlands, most common in low, swampy areas. They are often encountered about the breeding pools in early spring. At other times of the year they are secretive, but are sometimes found beneath logs and stones in wooded areas.

Kentucky Distribution: Poorly known. Breeding colonies are known in Bullitt and Meade counties; a single specimen has been taken in Carter County and another at Mammoth Cave.

Life History: Little is known of the life history of this species in Kentucky. In New York, where it is common, the sexes congregate in the breeding pools in March and April. Each female lays a few hundred eggs in clusters of from 7 to 40 attached to twigs or other support beneath the surface of the water. The clusters are normally cylindrical and usually about 1 x 3 inches but are highly variable in size. The individual eggs are spherical and about 1/10 inch in diameter. Each has its own distinct envelopes surrounding it, and the entire egg mass is surrounded by a soft jellylike substance, forming a common envelope. Incubation period is about 1 to 1½ months. At hatching the larvae average slightly over ½ inch in length. They are strongly pigmented, the ground color greenish to brown

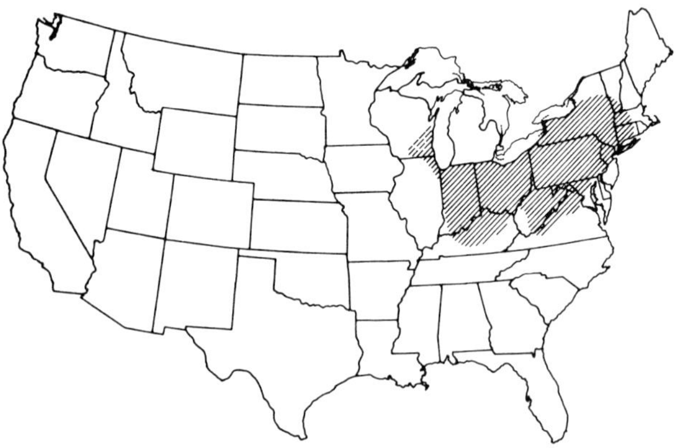

Range of Jefferson's Salamander (northern and western limits poorly known)

with tinges of yellow on the sides of the head and neck. There is a row of pigmented areas along either side of the mid-dorsal line, forming a series of conspicuous paired black spots down the back. The length of the larval period is variable, metamorphosis occurring from July through September. Youngsters begin breeding in their second spring, when nearly 2 years old.

Remarks: The first Kentucky specimen of this species was found on February 27, 1935, at Mammoth Cave National Park when a large stock pond was drained during road construction. Another specimen was taken in April from beneath a stone on a wooded hillside in Carter County. Diligent search of woodland pools in late February and March will likely reveal additional individuals.

Triploid Jefferson's Salamander II

Ambystoma platineum Cope

Recognition: This species cannot reliably be separated in the field from Jefferson's salamander. Only females are known and they usually have larger bluish flecks and spots than female Jefferson's salamanders. They can be positively recognized only by counting the chromosomes (42) or measuring the flattened surface of the erythrocytes (920-1510 square micra).

Variation: Other than individual variations in size, color, and pattern, none likely exists. Certainly, none is now known.

Confusing Species: Externally almost exactly identical to Jefferson's salamander, but the latter has 28 chromosomes and smaller erythrocytes (600-860 square micra).

Habitat: Woodlands. It is rarely encountered except in the breeding pools. Occasionally found hiding beneath a log or stone.

Range of the Triploid Jefferson's Salamander (northern and western limits poorly known)

Kentucky Distribution: Known only from Boone County in northern Kentucky. The distribution probably coincides with that of Jefferson's salamander.

Life History: Nothing is known of the life history of this species in Kentucky. Farther north, the eggs are laid in shallow woodland pools in March and April; they are deposited singly or in small clusters of 15 to 20 attached to leaves, sticks, or other debris on the pool bottom.

Remarks: This peculiar species is almost surely a fertile hybrid between Jefferson's salamander and the more northern blue-spotted salamander *(Ambystoma laterale)*. It occurs with Jefferson's salamander as a triploid population, apparently all females. They mate with male Jefferson's salamanders *(Ambystoma jeffersonianum)* and only triploid females are known to be produced. Thus, they are maintained. This is a rare, but not unknown phenomenon. Some question exists as to whether such organisms constitute a species. Although there are valid arguments both ways, they are here treated as a separate species.

A parallel situation exists between the blue-spotted salamander *(Ambystoma laterale)* and the triploid blue-spotted salamander *(Ambystoma tremblayi)* of the northern United States.

Small-mouthed Salamander
Ambystoma texanum (Matthes)

Recognition: A dark salamander with short legs and toes, a small, short, rounded head with a small mouth and small, pig-like eyes. Length usually 4 to 6 inches, to about 9½. Ground color dark gray, brown, or almost black, with pale irregular flecks over sides and back. The amount of light flecking on the body is highly variable, as is the ground color; some have so much light flecking, especially on the sides, as to appear rather silvery.

Variation: Apparently, the variations in color and pattern are not correlated with geography. There are, however, behavioral differences across the state (see below). Males generally have longer tails than females, and during the breeding season can be distinguished by their swollen vents.

Confusing Species: Most similar to the Jefferson's salamander complex; those two species have longer toes and a larger, longer snout and head.

Habitat: Most often encountered during February, March, and April in and along streams and bayous. As soon as the breeding season ends it leaves the water and goes underground in wooded or bushy areas, where it is sometimes found beneath logs, stones, or piles of debris.

Kentucky Distribution: Statewide except perhaps absent in the southeastern mountain counties. This species seems to reach its maximum abundance in Kentucky in the Inner Blue Grass, with the western portion of the state a close second. Although it may be encountered in the Eastern Coal Field, it seems much less abundant there than elsewhere.

Life History: In late January and early February they may be found in numbers in streams, ponds, and springs where mating occurs and eggs are laid. After mating the males apparently

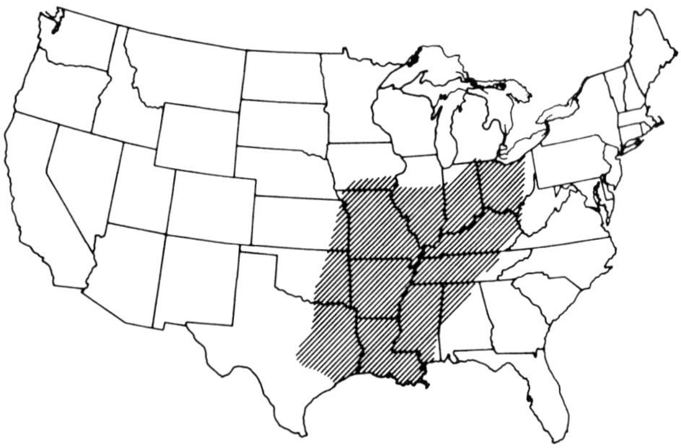

Range of the Small-mouthed Salamander

leave the water and go to higher ground, but the females remain near the water for several weeks. It is during the time that both sexes are in the water that they are most commonly encountered. They may sometimes be seen crawling along the bottom but more frequently may be found underneath a flat stone in a few inches of water.

Each female may lay up to about 700 eggs, either singly or in clusters of up to 25. The eggs are attached to submerged objects, at depths of up to about 6 inches. In eastern and central Kentucky the eggs are usually laid in streams; in western Kentucky they are laid in ponds and sloughs. A slanting stone with one edge out of the water and a clear space beneath it seems to be a favorite laying spot, at least in central Kentucky. The eggs are attached in a single layer to the lower surface of the stone, which sometimes shelters several hundred eggs.

Hard freezes frequently occur after the eggs are laid. Surely many of the shallower eggs are destroyed, but most of them seem to get by. At hatching the larvae are about ½ inch in length. In May or June, when about 2 to 3 inches long, they lose their gills, transform into adults, and leave the water.

Food of the adults apparently consists largely of earthworms, with an occasional insect.

Mole Salamander

Ambystoma talpoideum (Holbrook)

Recognition: A large-headed, dumpy, short-tailed salamander; somehow, it always looks deformed. Length usually 2½ to 4 inches, to about 4½. Color usually dark brown, but sometimes bluish or slaty, with assorted bluish or grayish flecks and spots. Often the dorsal crest of the tail is grayish. Costal grooves 10 to 11.

Variation: No geographic variation is known to occur. During the breeding season males may be identified by their swollen vents.

Confusing Species: Most like the two Jefferson's and the small-mouthed salamanders which, however, have 12 or more costal grooves and also have longer limbs and tails, slenderer bodies, and smaller, narrower heads.

Habitat: Poorly known, because of its burrowing habits. It is sometimes found beneath logs in swampy woodlands, but most often in the breeding pools.

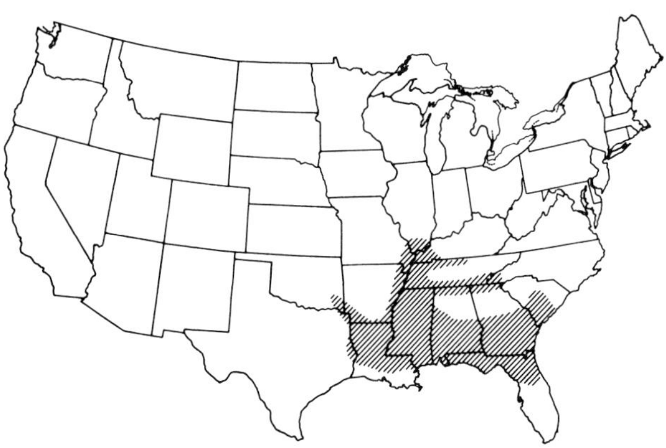

Range of the Mole Salamander

Kentucky Distribution: Poorly known. Western Kentucky, east at least into Todd County. Nowhere common, but most abundant in the lowlands of the Jackson Purchase.

Life History: Essentially nothing is known of the life history of this species in Kentucky. Individuals appear in the breeding pools in early spring. In Mississippi, they begin breeding in February; in Louisiana as early as January. The eggs are laid in the shallow water of temporary woodland ponds, in loose clusters of 4 to 20. Apparently the larvae transform in April or May, when something over half the adult size.

Marbled Salamander IV

Ambystoma opacum (Gravenhorst)

Recognition: A heavy-bodied black salamander with white bands across the back, often meeting on the sides to form an irregular white stripe. Length usually 3½ to 4½ inches, to about 5. One of the easiest of our salamanders to identify.

Variation: No geographic variation has been observed. Males are smaller than females and are more conspicuously marked. In females the contrast between the light and dark colors is subdued because the light areas are gray, rather than white.

Confusing Species: The color pattern is unique among our species.

Habitat: Occurs in a variety of habitats, from low swampy areas to relatively dry hillsides, but is essentially a woodland species. It is seldom encountered in the open, spending most of the time sheltered beneath logs, stones, boards, or debris.

Kentucky Distribution: Statewide. It seems to be most common in low-lying wooded areas of extreme western Kentucky,

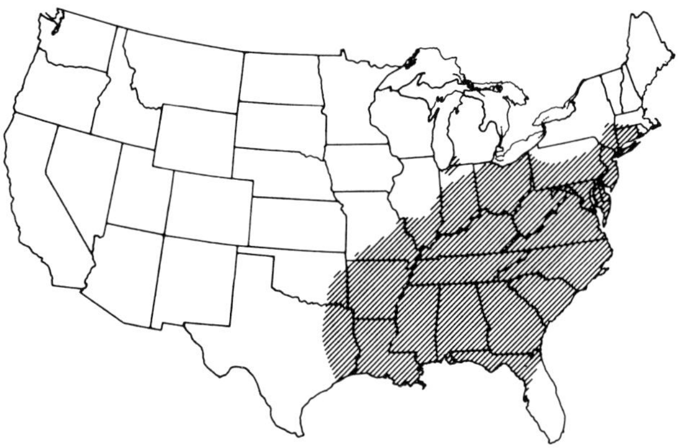

Range of the Marbled Salamander

less so in the mountains of eastern Kentucky, and least common in the areas in between.

Life History: Beginning about September, they leave their summer retreats and move into low-lying wooded areas that are likely to be flooded by fall rains. There the animals go through a sort of nose-rubbing nuptial dance (liebespiel) initiated by the males. Then the males deposit small packets of sperm atop little mounds of a jellylike substance they secrete. The females crawl over these packets and take them into the cloaca; the eggs subsequently are fertilized internally. A little later the female selects a depression in the ground beneath some type of shelter —a log, a piece of bark, moss, a pile of weeds—in an area likely to be flooded by the coming rains, and deposits about 100 eggs in it. She coils about the egg mass, and normally remains with it until the area is flooded. The eggs do not hatch until they are under water, so the incubation period is widely variable. They ordinarily hatch in the fall but may not hatch until the following spring.

Under normal conditions, the youngsters hatched in the fall metamorphose into adults and leave the water the following June or July. They go out into wooded areas and spend the summer hidden away under some sort of cover, often logs and stones. The youngsters do not reach sexual maturity the year they metamorphose and do not join the breeding aggregations until their second fall, when they are about 15 months old.

Food consists of a wide variety of small animals, but largely of assorted insects and worms.

Spotted Salamander V

Ambystoma maculatum (Shaw)

Recognition: A large, blue-black salamander with 2 irregular rows of rounded yellowish spots down the back. Length usually 6 to 8 inches, to about 9. Often the spots just back of the head are orange. The belly is gray.

Variation: Remarkably uniform across the state. There is some variability in the amount of orange in the anterior dorsal spots, but it is not certain that this is correlated with geography. During the breeding season males may be recognized by their greatly swollen vents and darker bellies.

Confusing Species: Most similar to the tiger salamander, which has more irregular, less rounded spots extending farther down on the sides, and a yellowish-olive belly.

Habitat: Most frequently encountered during the breeding season in late winter and early spring in or about woodland pools. They are seldom seen at other seasons because they normally remain hidden beneath logs, rocks, or piles of debris in woodland areas.

Kentucky Distribution: Statewide. Like many amphibians it does not occur or is uncommonly rare in certain areas. Despite many years of looking, I have yet to encounter a spotted salamander in the Inner Blue Grass. It is fairly common in the

Eastern Coal Field and quite common in western Kentucky. Apparently one requirement for this species is the presence of woodland pools. These simply do not exist, except in isolated instances, in the Inner Blue Grass, and various other areas of the state are notably shy of woodland pools.

Life History: In late February or early March spotted salamanders congregate in woodland pools to breed. They are voiceless, of course, but they do go through a series of nose-rubbings and assorted gyrations that constitute courtship. The males deposit little packets of sperm atop cone-shaped gelatinous mounds which they secrete on the bottom of the pool; the females swim over these packets and insert them into the cloaca.

A few days later the females lay their eggs in masses, varying from only a few up to 250 or so eggs. The average number of eggs per mass is about 125, and a typical egg mass is about 2½ to 3½ inches in diameter and is surprisingly compact. Obviously, a 6-inch salamander could not possibly contain such a monstrous mass. The explanation is that the eggs are covered with gelatinous envelopes that take up water and swell to their large size after the eggs are laid.

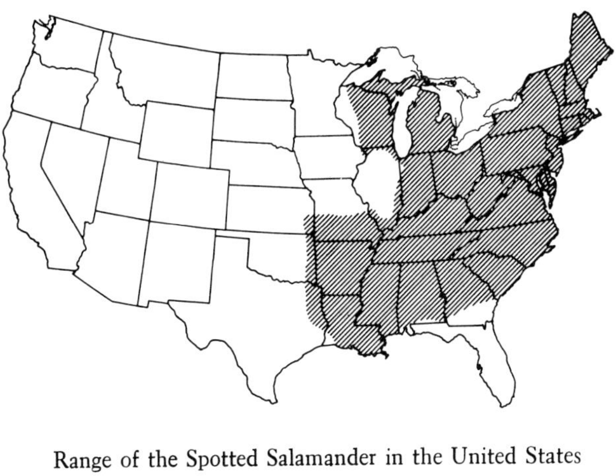

Range of the Spotted Salamander in the United States

Hatching occurs in about a month or two, depending on temperature. The larvae at hatching are about ½ inch in length, dull greenish yellow, with various darker spots scattered over the head and body. In about two or three months, the youngsters have attained a length of about 2½ to 3 inches and are ready to metamorphose into adults. Metamorphosis usually occurs in late August, September, or early October and they acquire the characteristic yellow spots of the adult within a few days. They grow to about 4 inches their second summer and are sexually mature the next year.

Tiger Salamander V

Ambystoma tigrinum (Green)

Recognition: A large, dark brown to blue-black salamander, conspicuously marked with irregularly shaped olive to yellowish-brown blotches over the back and extending well down on the sides. Length usually 6 to 8 inches, to 13. The belly is greenish-yellow, with darker mottling. Costal grooves 12.

Variation: Although there is much individual variation in the dorsal markings, there is no evidence of geographic variation across Kentucky. The subspecies occurring in the state is *Ambystoma tigrinum tigrinum* (Green), the eastern tiger salamander. Males have longer tails and longer vents than females. During the breeding season a male may be recognized by his swollen vent.

Confusing Species: Most likely to be confused with the spotted salamander, which has rounded dorsal spots arranged in two irregular rows and a grayish belly.

Habitat: A confirmed burrower, this species can survive in a wide variety of habitats ranging from forested mountains through open grassland to swamps and marshes. It often utilizes crayfish burrows as shelters.

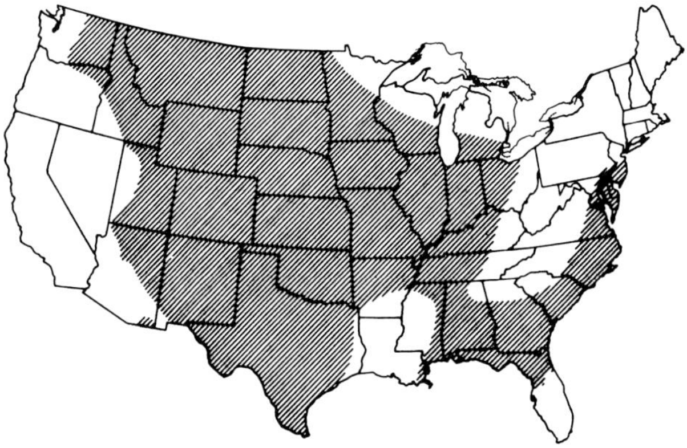

Range of the Tiger Salamander in the United States

Kentucky Distribution: Apparently absent from the Eastern Coal Field, but becoming increasingly abundant westward. It is common in some sections of western Kentucky, although not often encountered except in the breeding pools in early spring.

Life History: With the coming of the fall rains many tiger salamanders emerge from their burrows and may be found in the open. Perhaps this is an autumnal migration to the breeding pools, but breeding apparently does not occur until spring.

In February and March the adults congregate about the breeding pools. They usually select deeper water than other ambystomids and sometimes use farm ponds.

They engage in a pushing, nose-rubbing sort of courtship dance, followed by the deposition of spermatophores. The females pick up the sperm packets with the cloacal lips, and after fertilization the eggs are attached to underwater supports such as twigs in loose clusters of up to about 100. The egg mass with its encircling envelope is of variable size and shape, usually globular, elongate, or roughly rectangular, and measures from 2 x 2 to 3 x 4 inches. Hatching occurs in about 3 weeks, depending on temperature. The newly hatched larvae are nearly ¾ inch in length, greenish-yellow, with about 6 pairs of dark spots down the back. They usually metamorphose in July, at a length of about 4 to 5 inches.

Food consists of a wide variety of animal life, essentially anything that can be caught, subdued, and swallowed. Captive specimens readily take earthworms and ground beef.

FAMILY SALAMANDRIDAE Newts

A widespread family, occupying eastern Asia, Europe, North Africa, and North America. There are about 40 species, most of them European. Two genera occur in the United States, one in the east (*Notophthalmus*) and the other in the west (*Taricha*).

Eggs are laid in water, and the hatchlings go through an aquatic larval stage. Metamorphosis is usually complete, and the adults normally spend at least a part of their lives on land. The family is represented in Kentucky by a single species.

Newt VI-VII

Notophthalmus viridescens (Rafinesque)

Recognition: Adults usually have two quite dissimilar phases in the life cycle; however, in both they have a rather rough skin (compared to other salamanders) and lack costal grooves.

A terrestrial phase of from 1 to 3 years' duration usually follows metamorphosis. The terrestrial animal (red eft) is from 1½ to 3½ inches long, normally a bright orange-red, but sometimes yellowish. The skin is rough, and the tail is almost round in cross section.

In the aquatic phase the animal is from 3 to 5 inches long. Normally olive-green in color, it varies from yellowish to greenish brown. On occasion, as when forced onto land by a drying

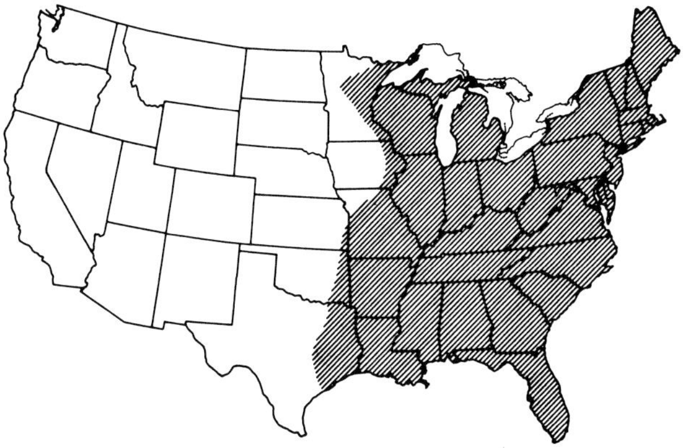

Range of the Newt in the United States

pond, it becomes almost black. The tail is much flattened, with a pronounced keel both above and below.

Variation: Two subspecies occur in Kentucky.

Notophthalmus viridescens viridescens (Rafinesque), the red-spotted newt, occurs from extreme eastern Kentucky across the state roughly to a line connecting Christian and Daviess counties. It is recognizable by the black-bordered red spots along the sides.

Notophthalmus viridescens louisianensis (Wolterstorff), the central newt, is the common newt of the Jackson Purchase. It is smaller than the red-spotted form, and lacks the black borders and usually even the red spots. This race seems more prone to metamorphose directly into the aquatic phase and omit the terrestrial phase than the red-spotted newt. The two subspecies intergrade in the area where they meet.

Males in the breeding season are readily recognizable by their high tail fins, swollen cloacae, and the black excresences on the inner surface of the hind legs. Females have horny, inward-turning margins on the cloacal lips. These disappear at the close of the spring breeding season, but may reappear by the following autumn.

Habitat: Red efts are most often encountered in moist woodlands, and the aquatic phase in woodland pools. They may use, however, nearly any quiet water, either in the woods or open. On occasion a farm pond may be occupied and harbor thousands of individuals.

Kentucky Distribution: Statewide. Everywhere common, but apparently least abundant in the Blue Grass.

Life History: In late winter or early spring the male performs an underwater nuptial dance in which he rubs the female with his cheeks and chin and then moves off a little distance and deposits a spermatophore. If his dance was successful, she follows and picks up the sperm-bearing cap of the spermatophore with her cloacal lips; fertilization is thus internal. A few hours later she begins to lay, often from 200 to 400 eggs over a period of several months. The eggs are laid singly, usually attached to a twig or stem, in quiet water.

After an incubation period of 3 to 5 weeks the eggs hatch; the newly hatched larva is slightly over ¼ inch long. Metamorphosis usually takes place in late summer or early fall, after a larval period of 2 to 3 months.

At metamorphosis the youngsters lose their gills, gain functional lungs, and usually develop the other characteristics of the red eft. Most leave the natal pool and take up a solitary terrestrial existence of from 1 to 3 years' duration. Some, however, metamorphose directly into the aquatic adult stage; these can often be recognized by their retention of the gill stubs.

At the end of the youngster's terrestrial wanderings, it returns to a quiet pool, enters the water, gradually acquires the appearance of the aquatic adult, and normally spends the remainder of its life in the water as a lung-breathing adult.

Remarks: No doubt the existence of two such dissimilar phases in the life cycle of this species has resulted in the plethora of names assigned the newt. Both the eft and the aquatic stage were given more than one name each before they were recognized as the same species.

FAMILY AMPHIUMIDAE Amphiumas

This is the smallest family of living salamanders, composed of a single species, *Amphiuma means,* the aquatic Amphiuma of the southern United States.

Amphiuma VIII

Amphiuma means Garden

Recognition: This big, dark-colored, aquatic salamander with its elongate, cylindrical body and tiny, obviously functionless limbs is easily recognized. Gills are lost at metamorphosis and do not occur in the adults.

The back and upper sides are dark brown to black and the lower sides and belly a light gray; the two colors meet along the sides and form a sharp line of separation. This is our longest salamander; an exceptionally large individual may reach nearly 3½ feet.

Variation: No geographic variation is known in Kentucky; *Amphiuma means tridactylum* Cuvier, the three-toed amphiuma, is the only form that occurs in state. The sexes can be distinguished externally by the shape of the vent. In the female it is a simple slit; in the male it is surrounded by a narrow raised ridge, and the inner surface toward the front is lined with oblique ridges.

Confusing Species: Superficially most like the true eel, a fish. The latter has internal gills, minute scales, and fins, rather than legs. Similar to *Siren,* which has external gills and only 2 legs.

Habitat: Most common in ditches, swamps, sloughs, cutoffs, and sluggish streams. In Kentucky, cypress swamps and oxbow lakes seem to be the preferred habitats. It may sometimes be encountered wandering overland in swampy areas on rainy nights.

Kentucky Distribution: Limited to the western area of the state. Apparently occurs in suitable habitats throughout the Jackson Purchase, but its eastern limits are unknown; it perhaps occurs as far east as Muhlenberg County, but it seems nowhere common.

Life History: Apparently this species spends the day coiled in or under a pile of debris under the water. Sometimes the head and a few inches of the body protrude from the hiding place, as if watching for a passing fish or crayfish, but often it remains completely concealed.

Amphiumas are not easy to catch. Not only are they extremely slippery, but they are also highly agile and will bite viciously when given opportunity. Stout gloves are advisable when attempting to capture a large one by hand.

The breeding cycle of Amphiuma is poorly known. Fertilization is apparently internal. Although essentially aquatic, the females forsake the water in late August, September, or early October to lay their eggs on land. A female selects (or roots

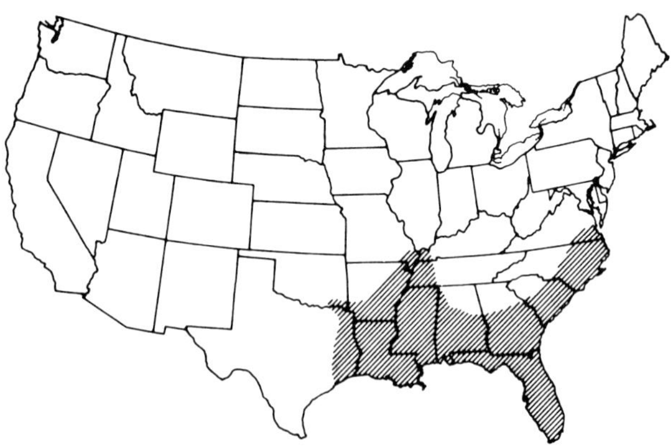

Range of the Amphiuma

out) a shallow depression under the shelter of a large log, stump, or similar debris, usually near the water, and deposits her eggs.

The eggs, numbering up to 150, are about ⅓ inch in diameter and are attached one to another like beads on a string. The clutch soon becomes twisted into a tangled mass; the female remains coiled about them until they hatch.

At hatching the youngsters are dark brown with scattered light spots and about 2½ to 3 inches long, of which roughly ½ inch is tail. They have short, light-colored external gills, and the feet and legs, although tiny, are well developed.

Food consists of a wide variety of animal life. Worms, insects, molluscs, and crustaceans are all eaten, as are salamanders, frogs, toads, fish, small snakes, and even an occasional smaller Amphiuma. Fishermen frequently take them on a baited hook in some areas.

FAMILY PLETHODONTIDAE Lungless Salamanders

This is by far the most successful family of salamanders. About 80 species, arranged in some 19 genera, are known. The distribution of the family shows some peculiarities; it occurs throughout most of North America and northwestern South America at least to Bolivia, and one genus with 2 species occurs on the Mediterranean island of Sardinia, the southeastern coast of France, and the adjacent northwestern coast of Italy. How these 2 species came to be there is unknown.

There seems little doubt that the family originated in what is now the eastern United States, almost surely in the southern Appalachians. These mountains to this day harbor more species, arranged in more genera, and more individuals than any other area. Kentucky with its eastern third stuck like a broad, jagged spear toward the heart of the southern Appalachians is abundantly supplied with plethodontids—7 genera, 16 species. Two species are represented by 2 subspecies each.

Adult plethodontids possess neither gills nor lungs; they derive their entire oxygen supply through the skin and mucous lining of the mouth. All exhibit a shallow groove (naso-labial groove) from the nostril down to the lip. In some cases the groove is small, and a lens is necessary to distinguish it. There is some question as to the function of the groove; perhaps it serves to drain and cleanse the nostril. On the other hand perhaps it is sensory, contributing to the sense of smell or taste or both. In some genera a pair of little projections (cirri) hang downward from the upper lip and carry the grooves to their tips. In any event, the presence of the naso-labial groove is a diagnostic feature of the family.

The method of attachment of the tongue to the floor of the mouth, while surely of little import to the salamander, is of value in identifying the animals. Some have the front edge of the tongue firmly attached to the floor of the mouth; others have it free all around, and attached by a central stalk, much like a tiny mushroom.

Two subfamilies are recognized. The more primitive of these, DESMOGNATHINAE, is represented in Kentucky by 3 species of the genus *Desmognathus*. These animals are peculiar in that the lower jaw is firmly attached to the skull, and is not at all movable. While an admirable adaptation for rooting about under stones, it poses some problems in opening the mouth. Rising to the occasion, the desmognathine opens his mouth by lifting the upper jaw and with it the top portion of the skull. This characteristic is reflected in the external appearance; the head, from a side view, seems set on the body at an odd angle.

The remainder of our plethodontids are in the subfamily PLETHODONTINAE and are characterized by a more prosaic lower jaw; it hinges downward.

Habitats occupied by plethodontids are quite diverse. Some are completely aquatic, some are stream bank inhabitants, and others have forsaken the water and gone into the woods and other moist places. Some are confirmed burrowers, and several species are cave dwellers, some white and blind, others not. One of our species (*Aneides aeneus*) has essentially left the ground to dwell in cracks and crevices of vertical cliffs or under the bark of dead trees.

Breeding habits are equally diverse, although fertilization is always internal. Some lay eggs on land, some in water. Some metamorphose before hatching, some after a larval period of varying length, and some never metamorphose at all, breeding in the larval stage. Some species (none in Kentucky) bear living young.

KEY TO THE GENERA AND SPECIES OF KENTUCKY PLETHODONTIDAE

1 a. Front edge of tongue attached to floor of mouth 2

b. Tongue free in front, attached by a central stalk 15

1a

1b

2 a. Five toes on each hind foot; tail without a constriction at its base; belly light or variously mottled, but never white with discrete black rounded spots 3

b. Four toes on each hind foot; tail with a basal constriction; belly white with discrete black spots / *Hemidactylium scutatum* (Schlegel), Four-toed Salamander (p. 72)

2b

3a

3 a. A light line from eye to angle of jaw *(Desmognathus)* 4

b. No light line from eye to angle of the jaw 8

4a

4b

4 a. Tail with a dorsal keel, trigonal in cross section 5

b. Tail not keeled above, round in cross section / *Desmognathus ochrophaeus ochrophaeus* Cope, Allegheny Mountain Salamander (p. 59)

5 a. Belly immaculate; dorsal pattern of paired light spots on

a darker background / *Desmognathus monticola monticola* Dunn, Appalachian Seal Salamander (p. 62)

b. Belly mottled; dorsal pattern various 6

5a

6 a. Size to about 6½ inches; dorsal pattern of numerous black blotches on a lighter background / *Desmognathus welteri* Barbour, Black Mountain Dusky Salamander (p. 57)

6a

b. Size to about 4½ inches; dorsal pattern various, often flecked with dark, but rarely with numerous black blotches on a lighter background (*Desmognathus fuscus*, Dusky Salamanders) 7

7 a. Adult dorsal pattern of 7 to 8 pairs of elongate yellow-

ish to red blotches; belly uniformly mottled; postocular bar broad, bright, yellowish or orange / *Desmognathus fuscus conanti* Rossman, Spotted Dusky Salamander (p. 55)

b. Adult dorsal pattern various, but usually not as above; body mottled, but usually with large unmottled areas; postocular bar narrow, whitish / *Desmognathus fuscus fuscus* (Rafinesque), Northern Dusky Salamander (p. 55)

8 a. With greenish blotches on body; tips of toes dilated, truncate; teeth large, nearly confined to front of mouth / *Aneides aeneus* (Cope and Packard), Green Salamander (p. 79)

b. Variable, but without green blotches on body; tips of toes rounded, and not dilated; teeth small, not confined to front of mouth *(Plethodon)* 9

9 a. With a wide dorsal band, which is lighter than the ground color, extending from head to near tip of tail 10

b. Without a bright dorsal band 12

10a

10 a. Dorsal band bordered by wavy or zig-zag dark bands / *Plethodon dorsalis dorsalis* (Cope), Zig-zag Salamander (p. 66)

b. Dorsal band not bordered by wavy or zig-zag bands 11

11 a. Costal grooves 20 to 23; dorsal band poorly developed, scarcely lighter than sides / *Plethodon richmondi richmondi* Netting and Mittleman, Ravine Salamander (p. 68)

b. Costal grooves 18 or 19, rarely 20; dorsal band well developed / *Plethodon cinereus cinereus* (Green), Red-backed Salamander (p. 63)

12 a. Belly as dark or darker than back, with few or no light flecks 13

b. Belly lighter than back, mottled with yellow or gray 14

13 a. Costal grooves 20 to 23; back, sides, and tail usually with minute gold flecks / *Plethodon richmondi richmondi* Netting and Mittleman, Ravine Salamander (p. 68)

b. Costal grooves 15 to 17; back, sides, and tail with many white spots / *Plethodon glutinosus glutinosus* (Green), Slimy Salamander (p. 70)

13a

13b

14 a. Costal grooves 19, rarely 18 or 20 / *Plethodon cinereus cinereus* (Green), Lead-backed Salamander (=dark phase of Red-backed Salamander) (p. 63)

b. Costal grooves usually 18, occasionally 17 or 19 / *Plethodon dorsalis dorsalis* (Cope), Zig-zag Salamander, dark phase (p. 66)

15 a. With a light line from eye to nostril, bordered below by a grayish line / *Gyrinophilus porphyriticus duryi* (Weller), Kentucky Spring Salamander (p. 74)

b. Without a line from eye to nostril 16

15a

16 a. Body slender, tail about as long or longer than body; naso-labial grooves often terminating below in small cirri *(Eurycea)* 17

b. Body stout, tail shorter than body; no cirri *(Pseudotriton)* 21

16a

16b

17 a. With a well-defined mid-dorsal stripe, or a near continuous series of elongated blotches forming a line down middle of back / *Eurycea longicauda guttolineata* (Holbrook), Three-lined Salamander (p. 84)

b. Without a well-defined mid-dorsal stripe or series of elongate blotches; at most a discontinuous series of small spots 18

17a

18a

18 a. With a definite dark band on either side of a light mid-dorsal band (*Eurycea bislineata,* Two-lined Salamanders) 19

b. Without a definite dark band on either side of a light dorsal band 20

19 a. Sides mottled, without light spots; males without prominent cirri / *Eurycea bislineata rivicola* Mittleman, Mid-west Two-lined Salamander (p. 81)

b. Sides nearly black, usually with light spots; males with prominent cirri / *Eurycea bislineata cirrigera* (Green), Southern Two-lined Salamander (p. 81)

20 a. With a definite dorsal band; sides of tail with vertical black bars; usually yellowish, but may be orange-red to red

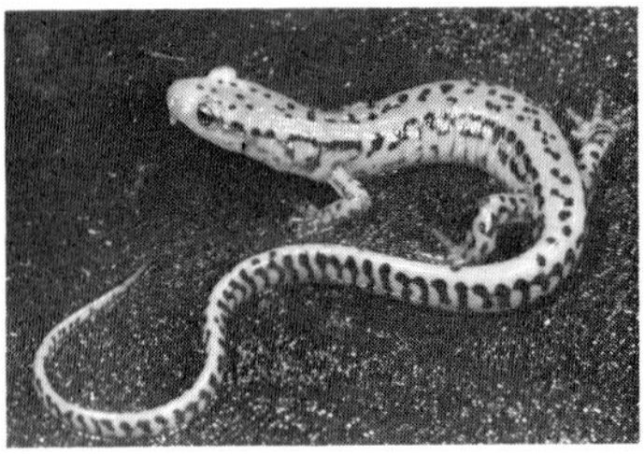

20a

20b

/ *Eurycea longicauda longicauda* (Green), Longtailed Salamander (p. 84)

b. Without a definite dorsal band; sides of tail with spots; usually reddish, but may be yellowish to orange / *Eurycea lucifuga* Rafinesque, Cave Salamander (p. 86)

21 a. Iris yellow; dorsal ground color bright coral-red to dark purplish brown; black spots irregularly circular, numerous, and tending to fuse in older individuals; vomerine series well separated at midline / *Pseudotriton ruber ruber* (Latreille), Northern Red Salamander (p. 77)

21a

21b

b. Iris brown; dorsal ground color light coral-red to brilliant red; dorsal spots nearly round, well separated, and almost never fused; vomerine series closely approximated at midline / *Pseudotriton montanus diastictus* Bishop, Midland Mud Salamander (p. 76)

Dusky Salamander

Desmognathus fuscus (Rafinesque)

Recognition: A dark stream-side salamander, usually readily recognized as a *Desmognathus* by the light line from eye to angle of jaw; this is sometimes obscured in old, very dark individuals. Highly variable in color, pattern, and size, with both age and locality. Youngsters show paired light spots down the back, but these are generally lost in the adult (See Spotted Dusky, below). Body rather robust, with hind legs somewhat larger and more muscular than fore legs. Length usually 2½ to 4 inches, to 4½.

Variation: Surely our most variable salamander; 2 subspecies are recognized in Kentucky.

Desmognathus fuscus fuscus (Rafinesque). Northern Dusky Salamander. Length to about 4½ inches. Dorsal pattern variable but frequently with a dark dorsal stripe. Belly mottled but usually with large unpigmented areas. Postocular bar narrow, whitish. Ranges from eastern Kentucky westward to about the western limits of the Western Coal Field.

Desmognathus fuscus conanti Rossman. Spotted Dusky Salamander. Length to about 4 inches. Dorsal pattern variable but usually with 7 or 8 pairs of elongate yellowish or reddish spots. Belly mottled. Postocular bar bright, wide, yellowish or orange. The common dusky salamander of the Jackson Purchase and the Land Between the Lakes. It intergrades with the northern dusky salamander roughly from the Cumberland River eastward to the Tradewater River.

Confusing Species: Most like *Desmognathus welteri* Barbour, the Black Mountain dusky, from which it differs by being considerably smaller and generally lacking dark dorsal markings. Similar to *Desmognathus monticola* Dunn, the seal salamander, but the latter has an immaculate belly.

Habitat: Although duskies may sometimes be found in moist woodlands, they are far more abundant under the stones and

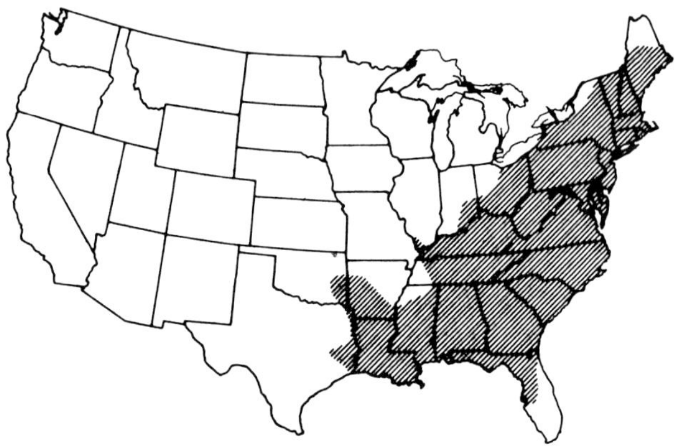

Range of the Dusky Salamander in the United States

logs along small woodland streams. They rarely venture more than a few yards from such a stream and then only in wet weather. Springs and spring runs are commonly inhabited. Nearly every little woodland stream in Kentucky supports a population.

Distribution: Statewide. They seem less common in the Jackson Purchase and in the Inner Blue Grass than elsewhere. They reach tremendously high populations in eastern Kentucky. Sometimes a little mountain brook, running only a trickle, shelters an adult dusky for every linear foot of stream.

Life History: This salamander is one of the few Kentucky species that can regularly be found in winter. Overturning some stones in a little flowing brook almost any day in winter will reveal a few active, but rather sluggish individuals.

They spend most of the daylight hours hidden away in a wet spot beneath a stone in the stream bed but are not at all reluctant to enter the water where they can travel at good speed. They are not difficult to catch once you learn the trick of herding one into an open palm with the other hand.

One summer 2 students and I kept track of the movements

of some dusky salamanders up and down a little stream in Breathitt County. We found a maximum movement of a little over 100 feet from May until October, and in this case the animal moved upstream as the lower, more open reaches gradually dried up. Where water was in continuous supply, movements on the order of 30 feet up and down the stream and up to about 5 feet out to the sides were about average for the 6 months. The animals seemed to know the good hiding places; each had a few preferred places in the stream whence it emerged to feed.

Eggs are laid in wet places, under shelter of a log, stone, moss, pile of leaves, or other debris from June to August. They are deposited in clusters of 12 to 30 and attended by the female who remains coiled about them. The larvae at hatching enter the water where they remain until metamorphosis. At this time, they are some 7 to 9 months old and about 1½ to 2 inches long.

Food consists of a variety of invertebrate animals, but mostly insects. Small crustaceans, millipedes and centipedes, spiders, and molluscs are also eaten.

Black Mountain Dusky Salamander IX

Desmognathus welteri Barbour

Recognition: A huge *Desmognathus*; color and pattern variable, but usually dark, with a dorsal pattern of black or dark brown blotches on a lighter ground color. Belly mottled. Length usually 3 to 5½ inches, to 6½.

Variation: No geographic variation has been observed. Males are usually larger, stockier, and have darker bellies than females. Vomerine teeth in males number from 3 to 6 and from 5 to 10 in females; males lose their vomerine teeth with increasing age. This species was originally described as a subspecies of *Desmognathus fuscus* and is still so treated by many authors. However, there is considerable evidence that it is a distinct species, and it is here so treated.

Confusing Species: Most like *Desmognathus fuscus*, but much larger and with larger dark dorsal blotches. Similar to *Desmognathus monticola*, but has a mottled belly and usually dark dorsal blotches on a lighter background while *D. monticola* has an immaculate belly and light dorsal blotches on a dark background.

Habitat: Basically a mountain brook form, most often found under stones in the stream bed, either in or near the water. Inhabits springs and spring runs and pools and roadside puddles in mountainous wooded areas.

Kentucky Distribution: The Eastern Coal Field, westward into the Knobs in the north, and westward across the Mississippian Plateau at least to Warren County. It reaches maximum abundance in the wooded streams of Big Black Mountain in Harlan County.

Life History: Little information is available on the life history of this species. Adults spend most of the day under the shelter

Range of the Black Mountain Dusky Salamander

of a stone or log in some little stream, emerging at night to feed. Fertilization is internal, and the female lays from 2 to 57 (average 26) eggs per season; larger females normally lay more eggs. Different individuals lay their eggs at different times. Egg laying begins in early March and accelerates gradually to a peak in August. By September a decline sets in, and by December all egg laying has ceased for the year.

The females remain coiled about the eggs until they hatch. A cluster of 16 eggs with an attendant female coiled about them was found in mid-June beneath a stone at the edge of a small stream in Harlan County. In mid-August a group of 14 larvae averaging slightly over ¾ inch in length was found with an attendant female in the hollow of a rotten limb at the edge of a mountain brook on Pine Mountain, also in Harlan County. Nothing is known of larval development.

Food is various, but consists largely of worms, insects, and crustaceans.

Remarks: The specimens upon which the scientific name of this species is based were collected near Lynch, in Harlan County.

Mountain Salamander x

Desmognathus ochrophaeus Cope

Recognition: A small, slender, round-tailed *Desmognathus*, with a straight-edged broad mid-dorsal stripe. It is the smallest member of the genus in the state. Length usually 2½ to 3½ inches, to 4. The color is extremely variable, although the pattern is relatively constant. The mid-dorsal band, running the length of the back and onto the tail, may be gray, brown, red, orange, or yellow. Frequently there is a series of small irregular blotches down the middle of the stripe. The sides are mottled, and the mottling fades imperceptibly into the lightly pigmented belly.

Some individuals are quite dark, and in these the dorsal stripe

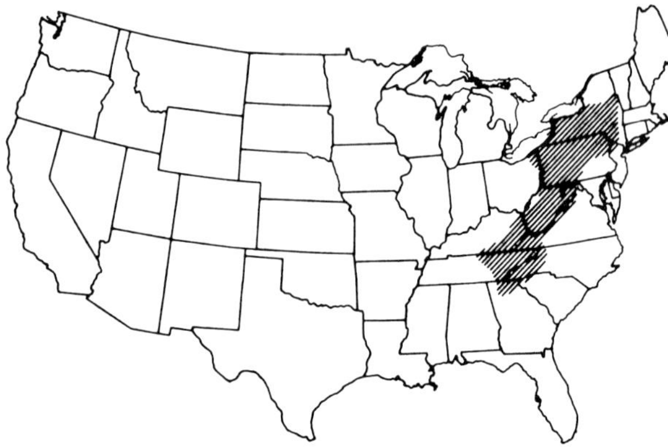

Range of the Mountain Salamander

may be obscured. Occasionally one is almost solid black with a brownish head.

In most individuals the light line from the eye to the angle of the jaw characteristic of all *Desmognathus* is well marked and quite obvious.

Variation: Little geographic variation has been observed in the state. *Desmognathus ochrophaeus ochrophaeus* Cope, the Allegheny Mountain salamander, is the only form that occurs. However, in Harlan County, roughly 5 percent of the population shows a tendency to the wavy-edged dorsal band characteristic of *Desmognathus ochrophaeus carolinensis* Dunn, the Blue Ridge Mountain salamander, a more eastern race. Males are somewhat larger than females and often become darker with increasing age. Some particularly large males are almost black, with little or no trace of the dorsal pattern remaining. Males lose the vomerine teeth with age.

Confusing Species: Most like the small species of *Plethodon*—zig-zag, red-backed, and ravine salamanders—all of which lack the light line from the eye to the angle of the jaw. Some specimens look surprisingly like the red-backed salamander.

Habitat: Moist woodlands, sometimes near streams, but often some distance away. All our other *Desmognathus* are stream and stream bank dwellers, but not this species. They may be found beneath rocks and logs on the forest floor, as well as in rotten logs. They are often found some distance off the ground beneath the bark of dead trees, stumps, or tree tops left from logging operations. On one occasion they were found in great abundance beneath stones in a brushy pasture a few yards from the woods in Harlan County.

On rainy nights in summer they can be found crawling over the forest floor or in little used roads and trails through the woods.

Kentucky Distribution: Throughout the Eastern Coal Field.

Life History: Although common, this animal is but infrequently seen. Most of the daylight hours are spent in hiding, and the animals usually emerge only at night or on dark, humid days.

Courtship and mating occur at night; a pair were observed engaged in a courtship dance (liebespiel) in Harlan County on July 14. After the courtship dance is completed the male crawls off to one side and deposits a spermatophore, a little stump-shaped mass of jelly bearing a whitish cap of sperm. If the dance was successful, the female picks up the sperm cap with her cloacal lips.

Eggs are laid in August, September, and October. There seems almost surely to be a spring period of sexual activity in April and May also. Eggs are deposited in little cavities in clusters of from 10 to 15 or so, or occasionally singly. They are white, with a diameter about that of a BB shot (4 mm.). The shelter of rocks, logs, clumps of moss, and the like in spring areas or near small streams are selected as nesting sites. The female remains coiled about her eggs until they hatch.

At hatching, the larvae are about ¾ inch long and have the dorsal band characteristic of the adult. They have short white gills, but these are soon lost.

Seal Salamander XI

Desmognathus monticola Dunn

Recognition: A large, robust *Desmognathus* with a plain, unmarked belly. Length usually 3 to 5 inches, to 5½. The dorsal pattern of the young consists of 4 or more pairs of light spots arranged down the back on opposite sides of the midline. In the adults these spots are much less conspicuous but are usually retained as a series of opposite or alternating lighter spots on the back, contrasting with the darker ground color.

Variation: No geographic variation in the state has been noted; the only subspecies occurring in the state is *Desmognathus monticola monticola* Dunn, the Appalachian seal salamander. Old males frequently lose the vomerine teeth, whereas the females retain them.

Confusing Species: Most like the dusky salamanders, from which it differs in having a plain, unmottled belly.

Habitat: Small streams, mountain brooks, and spring runs, usually in wooded areas. Sometimes they are found in rocky

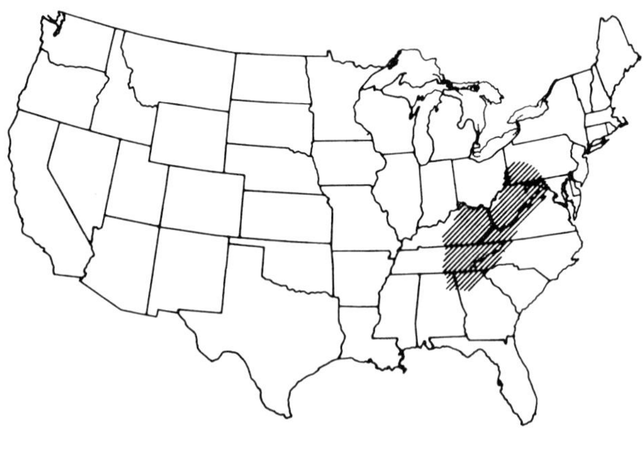

Range of the Seal Salamander

streams, but prefer muddier areas. An individual is frequently encountered with its head protruding from a burrow in a muddy bank along a small stream. When approached, it will usually withdraw into the burrow. If uncovered near water it does not hesitate to dive in and swim rapidly away.

Kentucky Distribution: Eastern Kentucky, westward in the north at least to Campbell County and at least to Warren County in the south. Most abundant in the east, but much less common than dusky salamanders in most areas.

Life History: Eggs are laid from July to September in clusters of from about 15 to 40. The female selects a sheltered spot near water, under a stone or log, or in a hollow of a rotten log. She apparently remains coiled about the eggs until they hatch.

The newly hatched youngsters are about ¾ inch long. They metamorphose when about 1½ to 2 inches. Food habits are poorly known, but they apparently feed on small invertebrates —worms, insects, crustaceans, and the like.

Remarks: For many years this animal was called *Desmognathus phoca*, and the specimen to which this name was applied was collected in Taylor's Creek, near Newport, Kentucky. *Phoca* is Latin for seal; when seen perched atop a wet rock at night, it does look somewhat like a miniature seal.

Red-backed Salamander XII

Plethodon cinereus (Green)

Recognition: A small, slender salamander with a straight-edged dorsal band, or sometimes uniform dark above. Length usually 2½ to 3½ inches, to 5. Two color phases occur:

1. Red-backed, with a straight-edged reddish stripe down the back from head to well onto the tail, bordered by dark pigment that extends down onto the sides.

2. Lead-backed, uniformly dark gray to nearly black, no dorsal stripe.

Intermediates are found, with the stripe present in varying degrees, or sometimes yellow, orange, or light gray in color. The belly is mottled black and white, giving a salt and pepper effect in both phases. Costal grooves 19, rarely 18 or 20.

Variation: Nearly every specimen yet examined from Kentucky is the red-backed phase. It is unlikely that any appreciable geographic variation occurs in the Kentucky population; males and females are externally similar.

Plethodon cinereus cinereus Green, the eastern red-backed salamander, is the only subspecies occurring in the state.

Confusing Species. The lead-backed phase is easily confused with Richmond's salamander, but the latter has 20 to 23 costal grooves and a nearly black belly. The red-backed phase may be confused with the zig-zag salamander, but the dorsal stripe of the latter has wavy or zig-zag edges. Similar to some *Desmognathus*, but these latter have a light line from the eye to the angle of the jaw.

Habitat: This is a completely terrestrial, woodland species, often found far from any permanent water. They are seldom encountered in the open and surely spend most of their time under logs, stones, or other sheltering objects in the woods.

Kentucky Distribution: The distribution of this species in Kentucky is poorly known. It occurs in Boone, Kenton, and Campbell counties and likely in several counties along the Ohio River to the east.

Life History: Red-backed salamanders are most commonly encountered under rocks and logs in early spring. Surely they must hibernate in such places, in crevices and holes extending beneath the frost line. As the season progresses and the earth becomes drier, the animals are less commonly seen; they un-

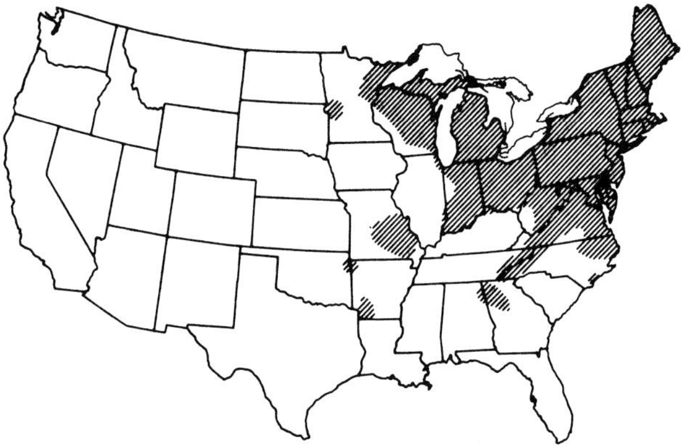

Range of the Red-backed Salamander in the United States

doubtedly go back again into the earth. They emerge during protracted, rainy spells and can then be found beneath rocks and logs. In the driest part of the year, in late summer and early fall, they remain secluded in moist areas in the earth and are rarely seen. With the advent of the fall rains, they are commonly encountered.

Nothing is known of the reproduction of this species in Kentucky. In New York, males deposit spermatophores on leaves and sticks in late fall, from October to December. The females recover the sperm packets and fertilization occurs. The eggs are laid the following year, in June and July. The female selects a small cavity, usually within a rotten log, sometimes under a stone, and deposits her usual complement of 8 to 10 eggs, hanging from the roof of the cavity by a common pedicel like a bunch of grapes. The female remains in attendance with the eggs, presumably until they hatch. Most of the eggs usually hatch in August. The newly hatched youngsters lose their gills within a few days; they do not go through a larval aquatic stage. The babies are well supplied with yolk and subsist upon it until able to shift for themselves. They reach sexual maturity when about 2 years old.

Food consists of a wide variety of small arthropods, annelids, and other animals.

Zig-zag Salamander

Plethodon dorsalis (Cope)

Recognition: A small, slender, salamander with a zig-zag or wavy-edged red dorsal stripe that continues broadly onto the tail. Length usually 2½ to 3½ inches, to 4½. Although usually red or reddish, the dorsal stripe may sometime be yellowish or silvery. In some individuals it is absent, often replaced by silvery mottlings.

The belly is bluish-white, strongly mottled with gray. The throat is lighter, and there is usually a light area, often red or orange, in front of the forelimbs. Costal grooves 17 to 19.

Variation: This species is quite variable in color and pattern, although these variations are not clearly correlated with geography. The race occurring in Kentucky is *Plethodon dorsalis dorsalis* (Cope). There is evidence that the population in south-central Kentucky, centering apparently in the area of Adair County, is distinct. Most of the individuals from that area are the silvery phase. Sexual dimorphism is not well marked externally, but males have more prominent chin (mental) glands, naso-labial swellings, and a pair of free lips at the posterior end of the vent.

Confusing Species: Most like the red-backed salamander, which has a straight-edged dorsal stripe. Dark forms of the two species are quite confusing, but *P. cinereus* usually has one more costal groove. The dark phase resembles some individuals of *Desmognathus* but lacks the light line from eye to angle of the jaw of that genus.

Habitat: Rocky areas, especially in open woods. Thin flat rocks, on the surface or slightly embedded in the forest floor, are favored hiding places. In early spring they can sometimes be found in great abundance in the wet leaves accumulated at the base of rocky outcrops.

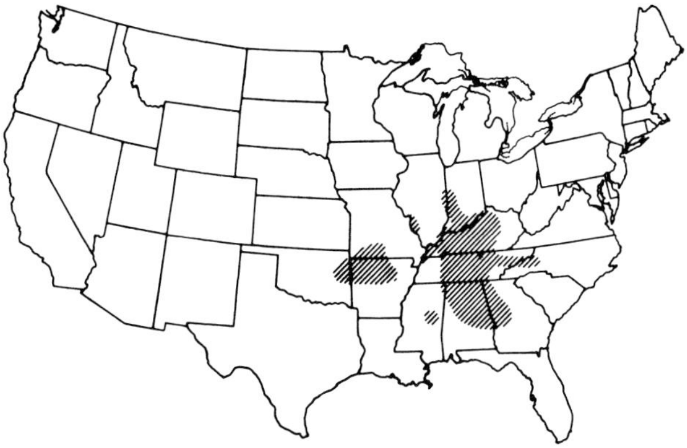

Range of the Zig-zag Salamander

Kentucky Distribution: From the western edge of the Eastern Coal Field westward across the remainder of Kentucky. This species is particularly abundant along rocky outcroppings in the Blue Grass and seems least common in the Jackson Purchase.

Life History: Surprisingly little is known about the habits of this abundant animal. They are rarely encountered in winter or summer but are easily found in spring and fall. Warm days as early as January will bring them near the surface, but they do not emerge from hibernation in great numbers until the advent of springlike days in March and April. With the coming of drier weather in May, they begin to retreat into the earth and seem not to emerge in numbers until the coming of the fall rains. Of course, a few individuals can always be found in moist places.

Remarks: The specimens upon which the scientific name of this species is based were collected in Louisville.

Ravine Salamander

Plethodon richmondi Netting and Mittleman

Recognition: A slender, dark, woodland salamander without conspicuous markings. The tail is long and rounded in cross section, the body long and slender, and the legs weak and set far apart—it looks almost like a worm with legs. Length usually 3 to 4½ inches, to 5½. Dorsal coloration brown to black, sprinkled with numerous minute gold or silvery-white flecks. Lower sides and belly dark, with many minute bluish-white flecks. Costal grooves 20 to 23.

Variation: Little geographic variation is recognized in Kentucky; *Plethodon richmondi richmondi* Netting and Mittleman is the only subspecies occurring in the state. Some individuals exhibit an ill-defined brownish mid-dorsal band, but most do not. In an occasional individual pale reddish blotches are scattered over the back and sides.

Confusing Species: Most closely resembles dark zig-zag and red-backed salamanders, but both of these have fewer costal grooves and a relatively shorter tail. Similar to some individuals

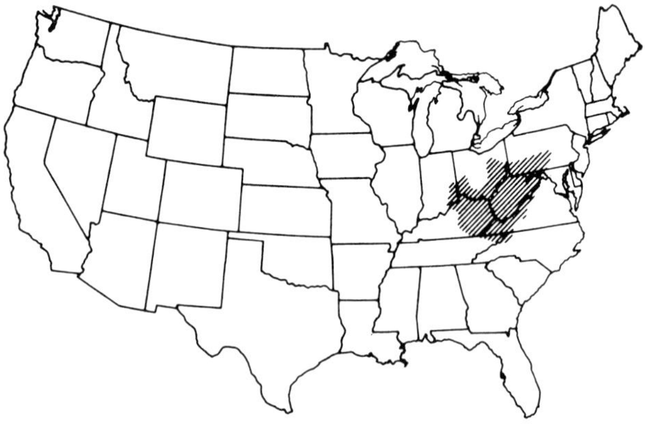

Range of the Ravine Salamander

of *Desmognathus ochrophaeus*, but these all have a light line from the eye to the angle of the jaw.

Habitat: A completely terrestrial species, most commonly encountered beneath thin flat rocks either on or but slightly embedded in the ground in open woodlands. They are most abundant on sloping hillsides but occur on the flats as well. They also inhabit dense, moist woodlands, taking refuge in the leaf mold on the forest floor, or beneath logs and stumps. Because of the greater abundance of suitable hiding places, they are more difficult to find in moist woods. They seem never to enter water voluntarily.

Kentucky Distribution: The Eastern Coal Field, the Blue Grass, and the Knobs. Ranges throughout eastern Kentucky westward to approximately a line from Jefferson to Whitley counties. It seems most common in the Blue Grass and the Knobs, but this is perhaps due to the fact that they can be more easily found in that kind of terrain.

Life History: In the moist woodlands of eastern Kentucky this species can be found throughout the spring, summer, and fall, although more commonly in April and May and in autumn after the fall rains begin.

In the Blue Grass and the Knobs they are rarely encountered in midsummer. They come up from deep in the ground in March or early April and may be found in numbers beneath flat stones on wooded hillsides. With the advent of drying days in May they begin to retreat deeper into the earth, apparently not to emerge again until autumn.

Breeding habits are poorly known. Two eggs, 2 newly hatched young, and 2 adults were found about 10 inches deep in the ground beneath a flat stone on a steeply sloping hillside in Mason County on August 23. One of the adults escaped into a crevice, but the other contained 4 ovarian eggs, each about 1/25 inch in diameter.

The hatchlings were light gray above, with an immaculate

belly. The eyes and eyelids were black and prominent. The dorsal pattern consisted of a large number of small round white spots in a network of black. The youngsters were nearly an inch in length and had presumably metamorphosed prior to hatching.

The 2 eggs ruptured when touched, and a youngster, indistinguishable from the two babies, emerged from each. They did not possess the larval characteristics of many newly hatched salamanders and were ready for a terrestrial existence.

Slimy Salamander XIII

Plethodon glutinosus (Green)

Recognition: A slender, moderate-sized black salamander with many small, irregularly spaced white flecks on the back, sides, and tail. The number of flecks is variable, sometimes only a few, sometimes so many, especially along the sides, as to make the animal appear to have white sides. Length usually 4 to 6½ inches, to about 7½. Costal grooves usually 16.

Variation: There is little geographic variation across the state; *Plethodon glutinosus glutinosus* (Green) is the only subspecies occurring in Kentucky. Males may be recognized by the presence of a rounded light spot, the mental gland, underneath the chin.

Confusing Species: Adults are distinctive. Half-grown young are similar to Richmond's and dark red-backed salamanders, which, however, lack white flecks and have more than 16 costal grooves.

Habitat: A woodland species. They normally spend the daylight hours beneath a rock or log, but at night and sometimes on rainy days they wander on the forest floor.

Kentucky Distribution: Statewide. They seem most abundant in eastern Kentucky, particularly in deep woods.

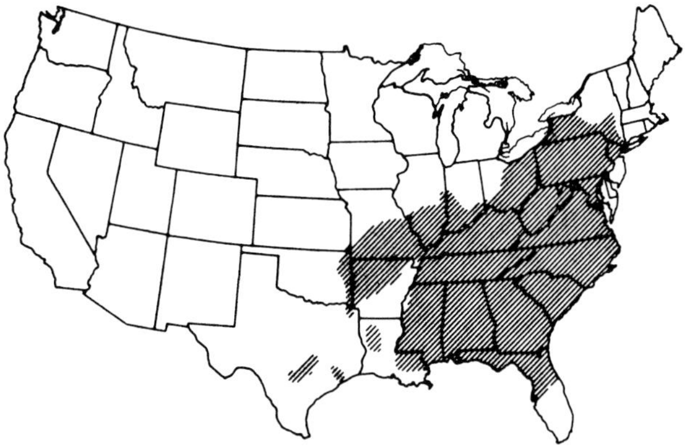

Range of the Slimy Salamander

Life History: With the coming of sunny days in March and April slimy salamanders come up from their winter retreats below the frost line and can be found under logs and stones in wooded areas. With the advent of the hot, dry days of summer, they apparently go deep in the ground in search of moisture, for they cannot survive drying. They do not often appear again until the rains of autumn, but a day or so of rain in midsummer is sufficient to bring them up again. In big timber, where there is sufficient organic matter on the forest floor to retain moisture, they may be found throughout the summer.

Surprisingly little is known of the breeding habits of this species. Apparently the female lays a cluster of 15 or so eggs in an underground retreat. Eggs with attendant females have been discovered from June to September. Incubation period is unknown. The eggs are not laid in water, and the young do not go through an aquatic stage; they pass through the gilled stage in the egg and metamorphose prior to hatching. Newly hatched babies look like miniature adults and are equipped to begin a terrestrial existence. Food consists of a variety of small animals, mostly insects and worms.

Remarks: The slimy salamander is one of the most appropriately named of all our salamanders. When handled, it exudes a whitish secretion from the skin that is quite slimy. As it dries the slime becomes sticky and when dry adheres to one's hands so tightly that only the most vigorous scrubbing will remove it. It will wear off in a few days, however. The scientific name of the animal in part reflects this characteristic; it translates roughly to "many toothed and bearing glue."

Four-toed Salamander XIV

Hemidactylium scutatum (Schlegel)

Recognition: A small, slender inconspicuous woodland salamander. Look for three unique characteristics in identifying this species: (1) belly enamel-white, with discrete, rounded, glossy black spots; (2) tail with a constriction at its base; (3) only four toes on each hind foot. Length usually 2½ to 3½ inches, to about 4. Dorsum reddish or yellowish brown, marked with darker spots or blotches.

Variation: No geographic variation has been observed. Males are usually smaller and slenderer than females, with longer tails and more truncate snouts. The vomerine teeth of the males are usually in a single row on each side, whereas females usually have 2 or more irregular rows on each side.

Confusing Species: Similar to several species of small woodland salamanders, but the three unique characteristics listed above will readily separate the four-toed salamander from all other Kentucky forms.

Habitat: A woodland species, sometimes encountered beneath rocks and logs or in moss. Larvae are aquatic, found in mossy or grassy shallow woodland pools, puddles, or small quiet streams. Adults are most frequently found in early spring when they congregate about such places to breed. Woods with nearby mossy or grassy pools are necessary for their existence.

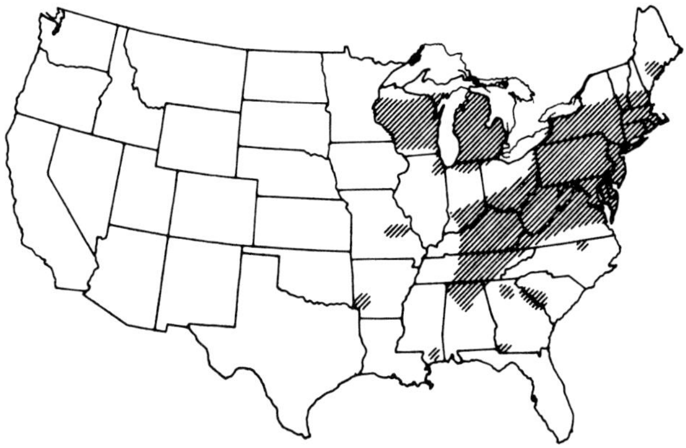

Range of the Four-toed Salamander

Kentucky Distribution: Poorly known; present records are quite spotty. Specimens are known from the counties of Bell, Breathitt, Bullitt, Carter, Edmonson, Jefferson, Madison, Menifee, Powell, and Wolfe. Probably occurs in isolated populations across Kentucky from the east westward at least to a line connecting Jefferson and Warren counties.

Life History: Mating occurs in the fall; fertilization is internal, by means of spermatophores. In March or early April the females emerge from hibernation and congregate about the breeding pools. Eggs are laid singly, in clutches of about 30, in cavities in moss or rootlets, usually within a few inches of the water. Clumps of moss or grass emerging from the water are favored sites. The eggs are glutinous and adhere to one another and to the surrounding moss and roots. The female remains coiled about the eggs until hatching.

Where the population is high, several females may lay eggs so close together that the clutches are indistinguishable; in such cases apparently only a few females remain with the eggs.

The eggs usually hatch in a month or so, but incubation may last as long as 2 months, depending in large part on temperature.

At hatching the larvae are about ½ inch long and unique

among plethodontids in the continuous keel running down the back and tail. They spend about 6 weeks as aquatic larvae and metamorphose at a length of an inch or slightly less. After metamorphosis they take up a terrestrial existence, returning to the water only to lay their eggs. They apparently do not reach sexual maturity until about 2½ years old.

Spring Salamander XV

Gyrinophilus porphyriticus (Green)

Recognition: A large, often pale, brownish pink to reddish salamander, with a light line, bordered below by a dark one, extending from the eye to the nostril. Length usually 4½ to 6 inches, to about 7. There are vague, dark dorsal spots and blotches usually arranged in a row along each side of the body, and the overall ground color seems veiled; it is not bright and clear. The belly is flesh-colored and unmarked, except for scattered black spots on the lower lip and occasionally the throat.

Variation: Only one subspecies, *Gyrinophilus porphyriticus duryi* (Weller), the Kentucky spring salamander, occurs in Kentucky. However, the population in extreme southeastern Kentucky seems to exhibit more black markings on the lower lip, chin, and throat than in other areas. Males are slightly larger than females.

Confusing Species: Most like the two species of *Pseudotriton*, but neither of them has a light line from eye to nostril. They also have brighter, less clouded coloration, and the dorsal spots are not usually arranged in a row along either side.

Habitat: Basically an inhabitant of a mountain brook or spring; most often encountered beneath stones in woodland streams. A favored site seems to be the shelter of a flat stone a foot or more across, lying at the edge of a clear woodland brook, partly in and partly out of the water. They sometimes are

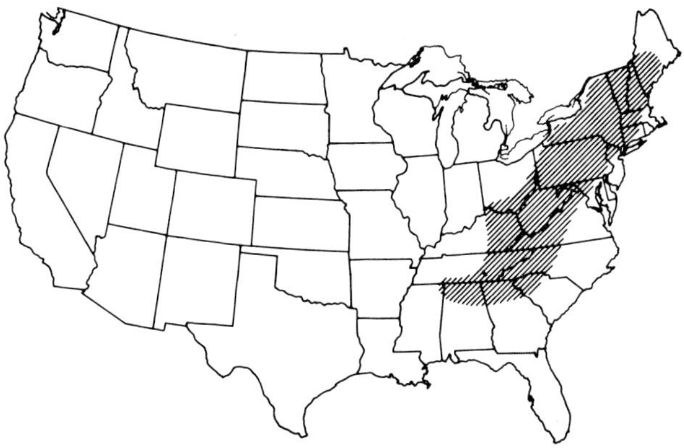

Range of the Spring Salamander in the United States

found at considerable distances from streams but seem always to be limited to wet places. A wet roadside ditch in a wooded area will sometimes harbor an individual or two. They often enter wet caves, going well back into the lightless zone.

Kentucky Distribution: Throughout the Eastern Coal Field. Western limits are unknown, but the species ranges at least into Fayette and Jessamine counties in the north and Monroe County in the south.

This species seems nowhere abundant, but diligent search along almost any clear, small woodland stream in eastern Kentucky will reveal an individual or two.

Life History: Essentially nothing is known of the life history of this animal in Kentucky. Its nearest relatives lay their eggs in water, attached to the underside of some support. They undergo a rather long larval period; larvae as long as 4 inches may sometimes be encountered. Some closely related species are neotenic, and this one may be.

Remarks: The specimen to which the name of this subspecies is assigned was collected at the base of a waterfall at Cascade Caverns in Carter County.

Mud Salamander

Pseudotriton montanus Baird

Recognition: A rather stout, short-tailed, brown-eyed, red salamander, sparingly sprinkled with discrete black spots. Size usually 3½ to 6 inches, to about 7. Color coral pink to red, sometimes brownish in older individuals. Spots few in number, and well separated. There is considerable individual variation in color and in number of dorsal spots. Older individuals tend to be darker, with the spots less conspicuous than in the young. Belly plain, unmarked in young, but may be flecked or mottled with brown in old individuals.

Variation: No geographic variation is recognized in Kentucky; *Pseudotriton montanus diastictus* Bishop, the midland mud salamander, is the only race occurring in the state.

Confusing Species: Most like the northern red salamander, but with many fewer black spots, and with a brown, rather than a yellow eye. Similar to the Kentucky spring salamander, but brighter, with more obvious spots, and without a light line from the eye to the nostril.

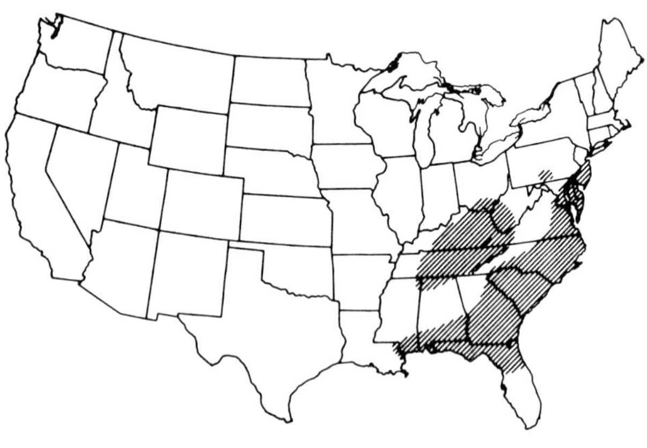

Range of the Mud Salamander

Habitat: Muddy areas. Most often encountered under stones or debris along a shallow, sluggish stream or spring run. Sometimes found in wet areas, such as roadside ditches, at a considerable distance from water. One individual was found under a stone in a dry, rocky pasture in Harlan County, many yards from any area that could be considered wet. Although apparently not limited to wooded areas, it seems more common along woodland streams.

Kentucky Distribution: Throughout eastern and central Kentucky; western limits of the range unknown, but extending at least into Trimble, Jefferson, Nelson, Edmonson, and Todd counties. This species seems nowhere abundant, but is probably as common in the Inner Blue Grass as anywhere; one can almost always find a few individuals in suitable habitats there.

Life History: Little is known of the breeding habits of this salamander. Apparently the eggs are laid in water, attached to submerged objects. At hatching the larvae have external gills, and most spend a period of time in the water. Size at metamorphosis is variable, ranging from about 2½ to nearly 3¾ inches.

Remarks: The specimen to which the name of this subspecies is assigned was collected at Cascade Caverns in Carter County.

Red Salamander XV

Pseudotriton ruber Sonnini

Recognition: A robust, short-tailed, yellow-eyed, reddish salamander, with numerous black spots scattered over the back and sides. Length usually 4 to 6 inches, to about 7½. Ground color red or red-orange, with numerous irregularly rounded black spots. Eye usually yellow. Old adults are darker, sometimes purplish brown, with the black spots tending to run

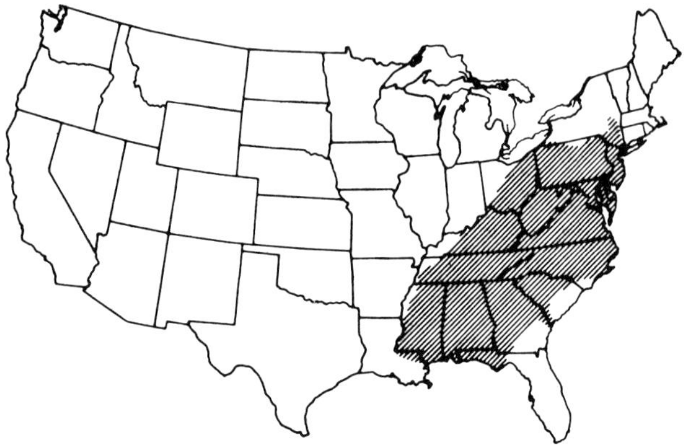

Range of the Red Salamander

together. Belly salmon-red, with small round brown or black dots; under surface of legs and tail usually unspotted.

Variation: No geographic variation is recognized in Kentucky; *Pseudotriton ruber ruber* Latreille, the northern red salamander, is the subspecies occurring in the state. External sexual differences are not well marked, but usually the anterior portion of the vent of the male is lined with low papillae, whereas that of the female is smooth.

Confusing Species: Most like the midland mud salamander, which has few, widely scattered spots, and brown eyes. Similar to the Kentucky spring salamander, but this animal has a light line from the eye to the nostril, and much smaller, less conspicuous dorsal spots.

Habitat: Clear, cold, rocky streams and springs, either in woods or open areas. Most often found under the shelter of a flat stone at or near the water's edge. Sometimes quite terrestrial, especially in summer, when it may be encountered under sheltering logs or stones in moist places at considerable distances from water.

Kentucky Distribution: Throughout the Eastern Coal Field, westward to and south around the Knobs at least to Meade and Breckenridge counties in the north and Edmonson and Todd counties in the south. Probably this species exists sparingly throughout the Blue Grass and the Knobs, but records are lacking.

Life History: The beginning of egg laying is apparently stimulated by the advent of the fall rains; most egg laying apparently occurs in October and November but sometimes later. The female selects a large stone a few inches under the water in a small stream under which to deposit her eggs. The individual egg is about the size of a BB shot (4 mm.), but including membranes is about 6 mm. in diameter. The eggs are attached individually to the bottom of the stone, and are arranged in small clusters. The total complement of eggs usually numbers about 6 dozen and may be several inches across.

Most eggs hatch in November or December, and the larvae are from ½ to ¾ of an inch in length at hatching. They apparently spend 2 years or more as an aquatic form before metamorphosing when about 3 to 4 inches long.

Food consists of earthworms, snails, slugs, spiders, centipedes and millipedes, and assorted other invertebrates. Occasionally a small salamander is swallowed.

Green Salamander XVII

Aneides aeneus (Cope and Packard)

Recognition: Our only green salamander. Length usually 3 to 5 inches, to about 5½. Body slender, flattened, tail round in cross section, toes truncate. Dorsum with green, lichenlike patches of color on a darker ground color.

Variation: No geographic variation has been observed. Males have fine papillae within the anterior lips of the vent. At least during the breeding season, the head of the male is somewhat swollen just behind the eyes.

Confusing Species: None.

Habitat: Cracks and crevices along the face of a sandstone cliff often harbor large numbers of green salamanders. Not all cliffs within the range of the species are suitable. There must be narrow cracks and holes for shelter. The cliff face must be moist, but not wet, and the openings must be at least partially sheltered from direct sun and rain.

Green salamanders occur, however, where there are no suitable cliffs; in such places they reside under the loose bark of dead trees and stumps. They sometimes reach a high population in logged-over areas where dead tree tops were left. They reached tremendous populations in the 1930s under the bark of the millions of dead chestnut trees in eastern Kentucky.

Kentucky Distribution: Throughout the Eastern Coal Field, and barely into the eastern Knobs. This species is particularly abundant in areas of standstone outcroppings, such as in Powell, Wolfe, and Menifee counties.

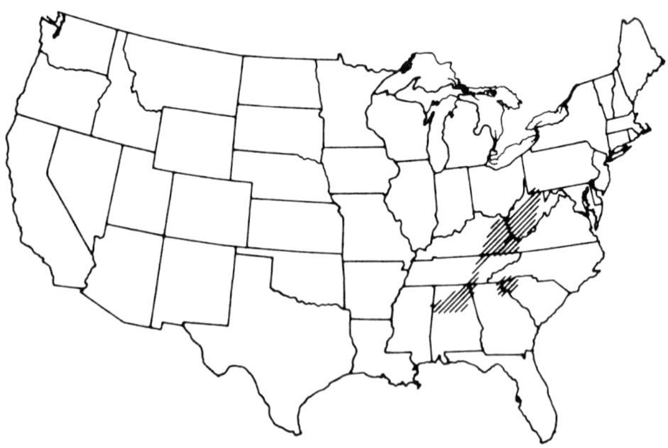

Range of the Green Salamander

Life History: The daylight hours are spent in hiding, and they emerge at night to feed. Mating habits are poorly known.

The eggs number from 10 to 20 and are usually laid in July or August. They are about the size of BB shot (4 mm.), laid in some sheltered spot where it is always moist and dark. The female stays with the eggs until they hatch and perhaps a little later, but there is little or no parental care of the young. The newly hatched babies look like miniature adults, even to color pattern, and are about ¾ inch long. As in most young, the proportions of the body are somewhat different from that of the adult. The nasal openings are quite large, the head is large, and the legs are small.

Two-lined Salamander XVIII

Eurycea bislineata (Green)

Recognition: A small, slender, yellowish salamander with a black line on either side of a broad dorsal stripe. Length usually 2½ to 4 inches, to about 4½. The dorsal stripe, often with brownish specks in it, is basically yellow but may be tinged with green, brown, or orange. The sides are usually mottled in varying degrees, sometimes plain, sometimes almost black.

Variation: This species varies considerably across Kentucky. The variation, long recognized, has resulted in at least two schools of thought concerning the status of the two-line in Kentucky.

According to one school, the subspecies *Eurycea bislineata rivicola* Mittleman, the midwest two-lined salamander, is the common two-line of Kentucky, ranging from the east across the state to the Jackson Purchase. In extreme western Kentucky it is replaced by the more southerly subspecies, *Eurycea bislineata cirrigera* (Green), the southern two-lined salamander.

The other school holds that the characteristics ascribed to *E. b. rivicola* fall within the range of variation of *Eurycea bislineata bislineata,* and the former is not a valid subspecies. If this be true, then the common two-line of Kentucky is *E. b. bislineata* (Green), the northern two-lined salamander. My own opinion is that *E. b. rivicola* is a valid subspecies.

Distinguishing characteristics of the three races have been suggested as follows:

Eurycea b. bislineata. Costal grooves 15 to 16, usually 15; adult males without pronounced cirri. Dorso-lateral black stripes usually not extending beyond the middle of the tail. Sides variously mottled, usually without light spots.

Eurycea b. rivicola. Costal grooves 14 to 16, usually 14; adult males without pronounced cirri. Dorso-lateral black stripes usually extending well beyond the middle of the tail, often reaching the tip. Sides mottled, often with light spots.

Eurycea b. cirrigera. Costal grooves 13 to 14; adult males with pronounced cirri. Dorso-lateral black stripes extending to the tail tip, or at least to its distal fourth. Sides dark, usually with light spots.

Confusing Species: Most like some individuals of *Desmognathus ochrophaeus,* but the latter have a light stripe from the eye to the angle of the jaw. Sometimes confused with the ground skink, *Lygosoma laterale,* a reptile, with scales, claws, and external ear openings. It might be confused with our other species of *Eurycea,* but adults of these all have a tail longer than the head and body and different patterns.

Habitat: Basically a brookside species, abundant along rocky brooks and small streams where it can often be seen perched atop a rock in a moist, shady spot. When disturbed it does not hesitate to enter the water and can swim rapidly by lateral undulations of the body. It is not limited to woodland streams, sometimes being found in numbers along streams flowing through open pastures.

In rainy weather it sometimes wanders considerable distances from open water and may subsequently be found hiding in moist places quite removed from streams.

Kentucky Distribution: Statewide. Nearly every little rocky stream in the state supports a population. It seems least abundant in the higher mountains of southeastern Kentucky and in the Jackson Purchase.

Life History: In late winter to early spring (January to April) the males prod and push the females with their noses until they become receptive to spermatophores. Beginning usually in April, eggs are laid singly, attached to the lower surface of some support (usually a rock) in running water. A complete complement numbers about 30 eggs, but several females may elect to lay under the same stone. The eggs, with their 2 surrounding membranes, are slightly larger than BB shot (4 mm.), and the outer membrane is drawn outward on one side to form an attachment disk.

Hatching occurs in August or September. The newly hatched larvae are ½ inch long; they have external gills and lack eyelids but resemble the adults in color and pattern. They metamorphose when 2 to 3 years old, usually at a length of about 2½ to 3 inches.

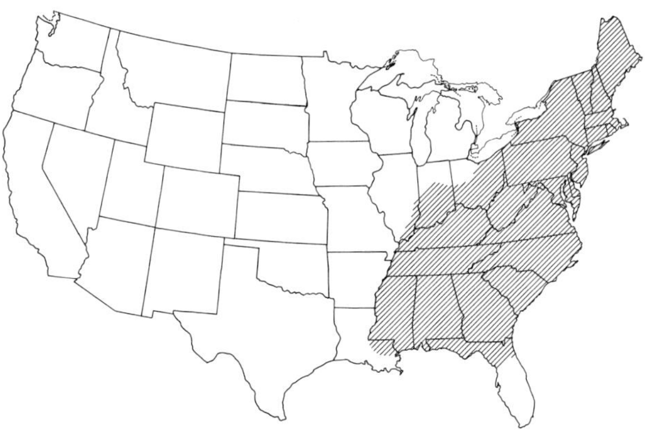

Range of the Two-lined Salamander in the United States

Long-tailed Salamander

Eurycea longicauda (Green)

Recognition: A long-tailed, slender, yellowish, speckled salamander with black vertical markings on the sides of the tail. Length usually 4 to 6 inches, to about 7¼. Color above generally yellow but may be greenish, orange-red, or sometimes red. Black markings on the body and sides variable in size and distribution but normally aggregated into dark vertical bars on the sides of the tail. Young are bright yellow and have relatively short tails.

Variation: This species exhibits wide variation across the state from east to west; 2 subspecies are known to occur. *Eurycea longicauda pernix* Mittleman has been recorded, but apparently this population is insufficiently well-marked or consistent to warrant subspecific status. The two subspecies currently recognized as occurring in Kentucky are:

Eurycea longicauda longicauda (Green), the Long-tailed Salamander. In this race the midline of the back bears at most scattered spots that do not form a mid-dorsal line. The spots on the sides are widely distributed, although sometimes so arranged as to suggest a dorso-lateral line. This subspecies ranges completely across the state except for approximately the southern half of the Jackson Purchase.

Eurycea longicauda guttolineata (Holbrook), the Three-lined Salamander. Similar to *E. l. longicauda*, but usually less yellow, more tan. Spots on each side aggregated to form dark dorsolateral stripes; dorsal spots aggregated in the midline of the back to form a continuous mid-dorsal stripe, or a series of elongate blotches. Apparently limited in Kentucky to the southern half of the Jackson Purchase.

Males differ externally from females in having a flange-like area around the vent and a few papillae on its inner surface. Both sexes sometimes develop cirri, but those of males are longer and slenderer.

Confusing Species: Most like *Eurycea lucifuga*, which lacks the vertical barring on the tail and is usually reddish, rather than yellow. Somewhat resembles *E. bislineata,* but in that species the tail is shorter and lacks vertical barring.

Habitat: The 2 subspecies exhibit different habitat preferences in Kentucky. The long-tailed seems particularly fond of moist or wet shale banks but is also found commonly under stones and debris along woodland streams. On occasion they wander away from the water and are sometimes found in moist places under stones or logs. The three-lined salamander seems partial to wet, mucky areas about springs and streams in river bottom areas.

Kentucky Distribution: Statewide, but it seems rarely abundant. However, Big Clifty Creek in Todd County supports a particularly dense population of the long-tailed salamander. The three-lined salamander seems common in the Blood River bottoms in Calloway County.

Life History: Little is known of the habits of this species. There is some evidence that eggs are laid in the fall, and incontrovertible evidence that they are laid in spring; likely there is a prolonged breeding season. The eggs are laid in water in clusters attached to the underside of a stone. The gilled,

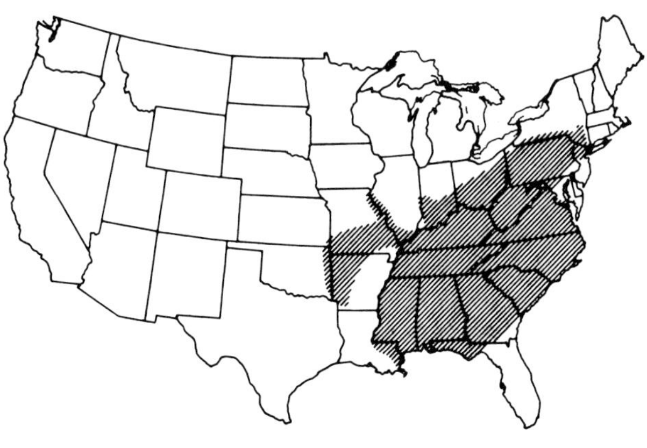

Range of the Long-tailed Salamander

aquatic larvae are about ¾ inch long at hatching and metamorphose when about 2½ to 3 inches long.

Food consists of various invertebrates—spiders, centipedes, millipedes, and many varieties of insects.

Cave Salamander XX

Eurycea lucifuga Rafinesque

Recognition: A long-tailed, slender, reddish salamander with many irregularly scattered black spots on the back and sides. Length usually 4 to 6 inches, to about 7¼. Although usually red, the ground color is variable, from yellowish through orange-red to red. Belly yellowish, usually unmarked. The young are normally yellowish, with shorter tails.

Variation: No geographic variation is recognized. On occasion the black spots are so arranged as to form 2, or sometimes 3, irregular rows and show considerable variation in size and frequency. Males have more swollen naso-labial areas and more prominent cirri than females.

Confusing Species: Most like long-tailed salamanders, which are usually yellow and have vertical black bars on the tail that the cave salamander lacks.

Habitat: Most frequently encountered in wet, limestone caves, where they are apparently more abundant in the dimly lighted zone near the entrance, but they can be found in the permanently dark zone. They also frequent areas far removed from known caves, taking refuge under stones or logs in wooded areas or along streams. They are most abundant in limestone areas, but are by no means confined there, occurring also in sandstone and granitic areas.

Kentucky Distribution: Statewide except the Jackson Purchase and a section of eastern Kentucky along Tug Fork,

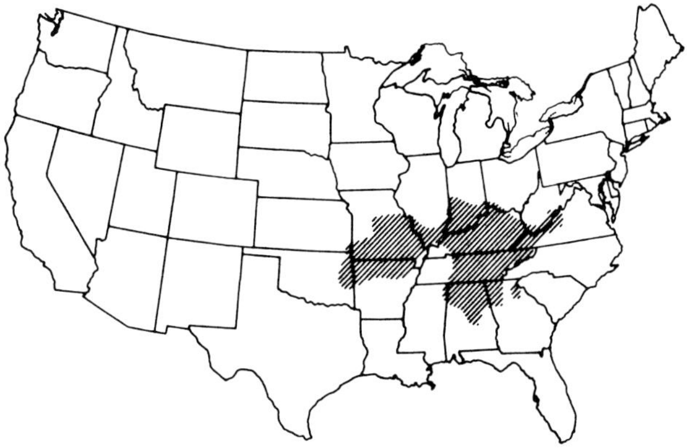

Range of the Cave Salamander

the Big Sandy, and the Ohio rivers roughly down to Maysville. However, it is common in the cave region of Carter County. The species seems most abundant in the Blue Grass, an area of many small, wet, limestone caves.

Life History: Until recently the life history of this species was essentially unknown, although the species was described in 1822.

Nocturnal, the cave salamander emerges from its hiding places and prowls about over the floor and walls of the twilight zone of caves or the forest floor with the coming of darkness. However, in the perpetually dark recesses of caves, they are liable to be active at any hour of the day or night. The prehensile tail is used in clambering about sometimes to support a dangling salamander. When disturbed, they leap wildly (several inches, at least) and show much agility in attempting to escape.

Courtship and mating have not been observed, but likely there is an elaborate courtship dance prior to mating. Eggs have been found from September 24 to November 5 in West Virginia and as late as January 20 in Missouri. Likely egg deposition in Kentucky occurs from September until January.

The eggs are deposited in water in caves; rimstone pools are known to be utilized, and there is some evidence that small sluggish streams in caves are also used. They must surely utilize other underground waters. The eggs usually are attached to the side of the pool by a stalk but sometimes they lie unattached on the bottom. They occur singly; rarely do two touch. The egg with its surrounding membranes is a little larger than a BB shot (4.1 to 4.4 mm.). A single female lays from 50 to 90 eggs, averaging about 70 per season.

At hatching the gilled, aquatic larvae are about ⅝ inch long; larvae approximately this size can be found in Kentucky caves in January and February. There is evidence that they are little more than 1 to 1¼ inches long when a year old. This slow growth is due in part, no doubt, to constant low temperature and shortage of food.

When the rimstone pools begin to overflow during the winter rains the larvae move out into the cave streams where they spend the remainder of their larval life. They metamorphose when about 2 to 2¼ inches long. When first transformed they are short-tailed and yellow, but the tail grows proportionally longer and the color deepens from pale yellow through deep yellow to red or red-orange with increasing age.

Remarks: The original specimens to which the scientific name of this species was applied were collected in early 1800s from caves near Lexington.

FAMILY PROTEIDAE Mudpuppies

This small family of permanently larval aquatic salamanders is composed of only 2 genera—*Necturus*, with 4 species in the eastern United States, one of which occurs in Kentucky; and *Proteus*, with one species of blind, white, cave salamander in Europe.

Fertilization is internal in the Proteidae, but little is known of courtship and mating. *Necturus* lays eggs, but ocasionally the eggs of *Proteus* are retained in the body of the mother to hatch there and then be born.

Mudpuppy

Necturus maculosus (Rafinesque)

Recognition: A large, dark, thoroughly aquatic salamander with bushy red external gills and 4 toes on each hind foot. Length usually 8 to 12 inches, to about 18. Color variable, but usually gray to dark brown, with blue-black rounded spots scattered over sides and back. Spots sometimes arranged in 2 irregular rows along the back, occasionally in a lateral row. Belly variable, sometimes light, sometimes pigmented, spotted or not.

Variation: Little geographic variation is recognized in Kentucky; *Necturus maculosus maculosus* (Rafinesque), the northern mudpuppy, is the subspecies in the state. However, the population in the Jackson Purchase tends toward *Necturus m. louisianensis*, the Red River mudpuppy. In males the vent is a longitudinal slit with wrinkled edges crossed behind by a groove, with two backwardly directed papillae. In the female, the vent is a simple slit.

Confusing Species: Sometimes confused with the hellbender, but the mudpuppy has external gills and lacks the fleshy fold

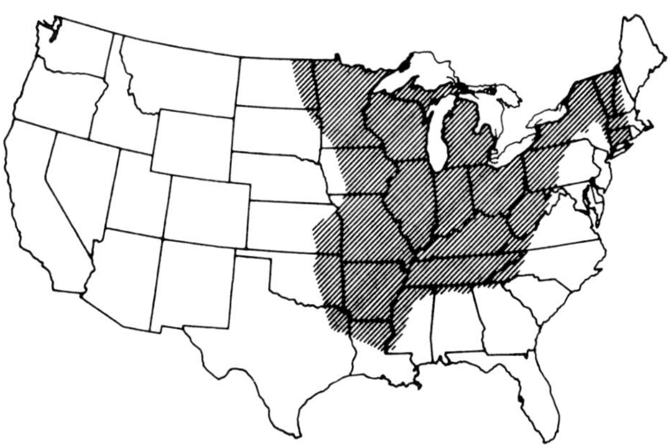

Range of the Mudpuppy in the United States

of skin along each side. Large larval *Ambystoma* are similar to small mudpuppies, but all *Ambystoma* have 5 toes on each hind foot.

Habitat: Thoroughly aquatic. Absent in mountain brooks and small streams, but occupies most streams sufficiently large to support a population of bass and other game fish. Inhabits lakes, impoundments and weedy, sluggish bayous, cutoffs, and drainage ditches. Frequently taken on trot lines, in fish traps, and by still fishermen.

Kentucky Distribution: Statewide; probably occurs in every suitable stream in the state. Difficult to collect by usual salamander collecting techniques so that data on abundance are scarce. It certainly is abundant in the Licking River drainage, at least in Rowan and Bath counties.

Life History: The mudpuppy is active throughout the year but apparently spends the colder months in deeper water. In clear water they are nocturnal, spending the daylight hours in seclusion under a rock, log, or pile of driftwood and emerging at night to feed. In muddy, stained, or weed-filled water they are active by day as well as at night.

Mating occurs in the fall. The males deposit spermatophores on the stream bottom which the females pick up with the cloacal lips; fertilization is internal.

In May or June the female lays her eggs, numbering approximately 100. They are about ¼ inch in diameter and attached to the undersurface of a submerged stone or other shelter. The female either selects or digs a cavity beneath the shelter large enough to contain both her and the eggs; she remains with them until they hatch.

Incubation is about 40 to 60 days, dependent on temperature. The newly hatched larvae are usually slightly less than an inch in length and are markedly different in appearance from the adults. There is a dorsal stripe from the snout continuing nearly to the tail lip, bordered on either side by a narrow yellow band extending from the gills along the upper sides of the tail. Below this, the sides are dark.

Since this species remains permanently in a larval stage, no metamorphosis occurs. Sexual maturity is reached when the young are about 8 inches long and approximately 5 years old.

A variety of animals, both vertebrate and invertebrate are taken as food, including fishes, amphibians, worms, molluscs, insects, and crustaceans.

Remarks: The specimens to which the scientific name of the species is assigned were collected in the Ohio River in the early 1800s.

ORDER ANURA Frogs and Toads

Many vertebrate orders are difficult to characterize, and even more difficult to recognize, but not the Anura. With their short, dumpy bodies, enlarged hind legs, and neither neck nor tail, they resemble no other animals and can be immediately recognized as frogs or toads.

Frogs generally have a smooth moist skin, no paratoid glands, and move in leaps. Toads, on the other hand, have a warty, dry skin, enlarged paratoid glands, and hop from one place to another.

Although all our frogs and toads fertilize their eggs externally, lay them in water, and go through a period of larval life as a gilled, aquatic tadpole before metamorphosing into a miniature adult, not all Anura display these characteristics. *Ascaphus* practices internal fertilization. *Eleutherodactylus augusti* lays its eggs on land, where they hatch directly into terrestrial froglets. Eggs of *Pipa* are implanted in the back of the female, who carries them until they hatch—in some species, into tadpoles, in others, into tiny frogs. The male *Alytes* carries a string of from 20 to 60 eggs wound about his hind legs for several weeks; just prior to hatching, he deposits them in water where they hatch and the tadpoles develop. There are several other genera with peculiar habits that could be cited, but these few cases are sufficient to indicate some of the variations that occur.

Our anurans usually congregate about pools or puddles in spring or summer to mate and lay their eggs. The males do the calling (most calls are distinctive and easily recognized), and although the females can and do utter various vocal sounds on occasion, they are ordinarily silent. When mating, the male mounts the back of a female and clasps his forelimbs tightly about her shoulder area. As she lays her eggs, he deposits sperm over them. The incubation period is variable, from a day or so to a few weeks, and the resultant tadpoles spend from some 2 weeks to about 2 years in the water before metamorphosing into miniature adults.

About 2,200 species of anurans are known, arranged in 16 families. Five families, 7 genera, and 17 species occur in Kentucky.

KEY TO THE FAMILIES OF KENTUCKY ANURA

1 a. With one or two sharp-edged, flat digging spurs on each hind foot — 2

b. Without a sharp-edged digging spur on each hind foot — 3

2 a. With a single spur on each hind foot; paratoids small, inconspicuous; pupil of eye vertically elliptical / Family Pelobatidae, Spadefoot Toads (p. 94)

b. With two flat digging spurs on each hind foot; paratoids large, conspicuous; pupil of eye horizontally elliptical / Family Bufonidae, True Toads (p. 116)

3 a. Tips of toes dilated into rounded adhesive disks / Family Hylidae, Treefrogs (p. 122)

b. Tips of toes gently tapering to tip, not dilated into rounded disks — 4

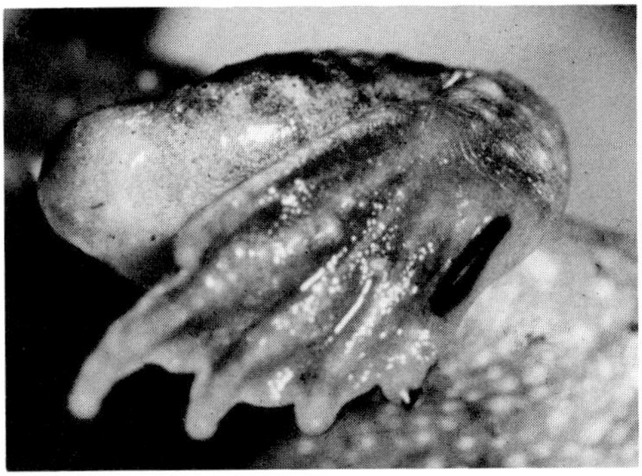

1a, 2a

2a

2b

3a

4a

4b

4 a. A transverse fold of skin across the head behind the eyes; feet not webbed / Family MICROHYLIDAE, Narrow-mouthed Toads (p. 114)

b. No transverse fold of skin across head behind eyes; hind feet extensively webbed / Family RANIDAE, True Frogs (p. 97)

FAMILY PELOBATIDAE Spadefoot Toads

Pelobatids are widely distributed in the northern hemisphere, and extend south of the equator in Indonesia.

The family is divided into 2 subfamilies. One, the Megophryinae, is confined to the Far East. This group includes some bizarre animals; the weird horned frog of Indonesia is a classic example. Each upper eyelid extends forward and upward to form a long, thin, pointed "horn," and there is a long, flexible projection of the snout; thus the animal looks as though it has 3 horns.

The spadefoot toads of Europe, North Africa, and North America constitute the subfamily Pelobatinae. They resemble the true toads (Bufonidae) in build, but have a smoother, less warty skin, and vertical pupils. The "spade" is a broad, sharp, flat-edged tubercle on the inner side of the hind foot with which these animals can dig themselves out of sight in short order.

The family is represented in North America by but a single genus, *Scaphiopus*, with 5 species. A single species occurs in Kentucky.

Spadefoot XXII

Scaphiopus holbrooki (Harlan)

Recognition: A toadlike animal with inconspicuous paratoid glands, few warts, and vertically elliptical pupils. Length, snout to vent, usually 2 to 2½ inches, to nearly 3. Usually brown, sometimes yellowish or quite dark, often mottled, with two light dorsal stripes usually; throat and breast white; belly tinged with gray. With a broad, flat-edged digging spur on each hind foot.

Variation: No geographic variation is recognized in Kentucky; *Scaphiopus holbrooki holbrooki* (Harlan), the eastern spadefoot, is the subspecies occurring in the state. Males have stouter fingers than females, with black horny deposits on the upper surface of the first and second ones. In the breeding season, males often have the throat dilated into a large vocal pouch.

Confusing Species: Most like toads of the genus *Bufo*, but they have conspicuous paratoids, many warts, horizontally elliptical pupils, and two flat-edged spurs on each hind foot.

Habitat: Terrestrial; subterranean. It spends most of the time in burrows which it digs, backwards, with the spurs on the hind feet. Though sometimes encountered abroad in wooded areas on dark, rainy summer days, the species is not limited to wooded areas. It has been uncovered by plows in cultivated fields. One individual is known to have burrowed over 2 feet deep in a pile of crushed limestone.

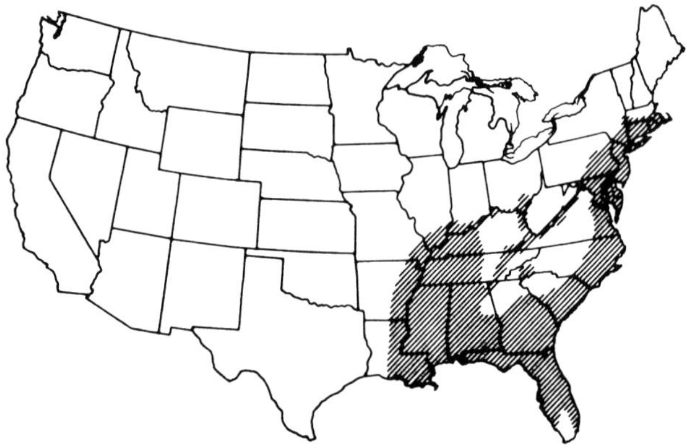

Range of the Spadefoot Toad

Kentucky Distribution: Poorly known. It has been found in Edmonson, Floyd, Hardin, Meade, Warren, and Whitley counties and likely occurs widely through the western half of the state, sporadically elsewhere. It is apparently absent from the Blue Grass and the Knobs. Although data are few, it appears most abundant in northwestern Warren County, in the vicinity of Hadley.

Voice: A deep, coarse, monosyllabic "wonk, wonk." A single calling male can be heard at least a mile and a large chorus is almost deafening. It normally calls only after periods of unusually heavy rain, from March to September, but in Kentucky perhaps most frequently in June. Calling is usually limited to the hours of darkness, but daytime choruses are known. The call is apparently used by males to attract a female, for the mated pairs are silent.

Life History: During or immediately after periods of unusually heavy rainfall in warm weather, the spadefoots emerge from their burrows and congregate about shallow, temporary pools to mate and lay their eggs.

When ready to lay, the female, with a male attached to

her back, sinks to the bottom of the breeding pool and selects a twig or blade of grass in the water. Pressing her vent close to it, she deposits her eggs in irregular bands along twigs or grass blades, 1 to 2 inches wide and up to a foot long. Each egg is about 1/16 inch in diameter, but including envelopes is about 3/16 to nearly ¼ inch. The male fertilizes the eggs as they are laid.

After an incubation period of only 1½ to 2 days, the eggs hatch into bronzy, broad, short-tailed tadpoles. Metamorphosis occurs from 2 to nearly 8 weeks after hatching, depending upon temperature and water level.

Remarks: The very short incubation and brief larval periods are apparently an adaptation to breeding in temporary pools and puddles. Breeding occurs only during a day or so immediately following an exceptionally heavy rain and does not occur at all in many years.

FAMILY RANIDAE True Frogs

Although the Ranidae have long been considered the most advanced family of our frogs and toads, substantial evidence now indicates that they are relatively primitive and belong near the bottom of the evolutionary sequence in the order Anura. Be that as it may, the family is highly successful in point of numbers and widespread distribution. Evidently the family originated in Africa, for 6 of 8 recognized subfamilies (9 by some authors) occur there, and 4 are found nowhere else. One subfamily occurs only in tropical America. The most successful subfamily is Raninae, comprising the genus *Rana* and a few close associates; the largest known frog, an African species, is a member of this subfamily. *Rana* is by far the most abundant and widespread genus, containing some 200 known species and found in all the major land masses except Greenland, New Zealand, central and southern Australia, and southern South America.

All 6 species of the family known in Kentucky are members of the genus *Rana*.

KEY TO THE SPECIES OF KENTUCKY RANIDAE

1 a. Without dorso-lateral folds on trunk; short folds present behind the eyes, but ending near the eardrum / *Rana catesbeiana* Shaw, Bullfrog (p. 101)

b. With an obvious ridge (dorso-lateral fold) of skin along each side of the back, extending from just behind the eye at least halfway to the hip 2

1a

1b, 2a

2 a. Dorso-lateral folds ending on body, not reaching groin (*Rana clamitans* Latreille, Green Frog) 3

b. Dorso-lateral folds extending to groin 4

3 a. Dorsal color green to greenish brown, with dark brown or gray-brown blotches; throat of male bright yellow; maximum snout-vent length over 3 inches; belly usually not heavily blotched or reticulate / *Rana clamitans melanota* Rafinesque, Green Frog (p. 103)

b. Dorsal color brown to bronze, without dark blotches; throat of males sometimes yellowish but usually dusky;

maximum length of snout-vent under 3 inches; belly usually heavily blotched or reticulate / *Rana clamitans clamitans* Latreille, Bronze Frog (p. 103)

4 a. With a large black patch on cheek, covering the lower half of the eye; dorsum usually with no conspicuous spots or blotches / *Rana sylvatica sylvatica* Le Conte, Eastern Wood Frog (p. 110)

b. Without a large black cheek patch; with conspicuous dorsal spots 5

4a

5a

5 a. No white line on upper jaw; dorsal ground color milky white, with many large, irregular dorsal spots, many with light centers / *Rana areolata circulosa* Rice, Northern Crawfish Frog (p. 112)

b. With a white line on upper jaw; dorsal ground color not milky white, spots more or less regular and regularly spaced 6

6 a. Under side of hind legs orange; two rows of squarish black spots between the dorso-lateral folds / *Rana palustris* Le Conte, Pickerel Frog (p. 107)

b. Under surface of hind legs white; dorsal spots rounded, light-edged, and arranged in 2 or 3 rows (*Rana pipiens* Schreber, Leopard Frog) 7

6a

6b

7 a. A white spot in center of the tympanum; a few dark spots on sides; head longer, more pointed / *Rana pipiens sphenocephala* Cope, Southern Leopard Frog (p. 105)

b. No white spot in center of the tympanum (may be light, however); numerous dark spots on sides; head shorter, more blunt / *Rana pipiens pipiens* Schreber, Northern Leopard Frog (p. 105)

Bullfrog

Rana catesbeiana Shaw

Recognition: A large frog without dorso-lateral folds on the trunk. Length, snout to vent, usually 3 to 6 inches, to 8. Dorsal color green, gray, brown, or black; plain, or marked with wormlike markings or blotches; belly whitish, often mottled with gray or having a yellowish cast.

Variation: No geographic variation is recognized. Males are smaller than females, and have larger tympani, swollen thumbs, yellow throats, and openings from the throat into the vocal sacs.

Confusing Species: Most like the green and bronze frogs, but both these have dorso-lateral folds on the trunk.

Habitat: Lakes, impoundments, farm ponds, large springs, rivers, creeks, and the larger drainage ditches, wherever water is permanent. They are normally replaced by the green frog in smaller tributaries, but occasionally a young bullfrog will be encountered in small tributaries almost to the point of the last permanent water.

Kentucky Distribution: Statewide; abundant nearly everywhere. Perhaps least abundant in the higher mountains of southeastern Kentucky.

Voice: The calls have great carrying power (a mile or so) and are perhaps best characterized as "jug-o-rum," or "br-wum." The males begin calling usually in late April or early May and continue until August and sometimes September. Calling becomes rather sporadic later in the season.

Life History: After emergence from the mud in spring the males establish a territory along the water's edge and defend it from other males. They are easily alarmed by day, leaping

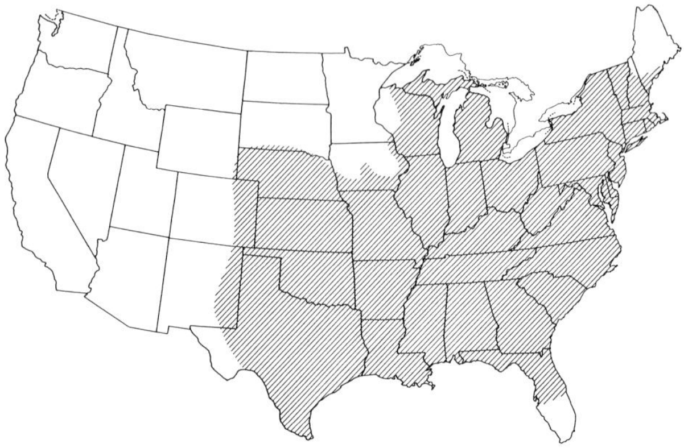

Range of the Bullfrog in the United States (also widely introduced in the western states)

into the water with a great splashing when disturbed, but can often be approached and captured by hand at night.

Mating occurs from May to July or August, and the eggs are laid in great floating mats, sometimes a yard or so across. A single female will lay upwards of 20,000 eggs a season, of which only a minute percentage reach maturity. The eggs hatch into tadpoles which lead a completely aquatic existence for about 2 years, sometimes reaching a length of nearly 6 inches. The young frogs after metamorphosis look like miniature adults, but 2 or 3 more years will elapse before they reach adult size. Thus 4 or 5 years are required from egg to adult.

Bullfrogs feed on a wide variety of animals. Beetles, moths, crayfish, crappies, sunfish, assorted species of frogs and toads, salamanders, snakes, turtles, birds, mice, and bats have been recovered from frog stomachs. No doubt a bullfrog will eat any animal it can capture, subdue, and swallow; however, insects and crustaceans make up the bulk of the diet.

Remarks: The bullfrog has been introduced in many parts of the world, particularly in the Far East, as a possible source of additional protein. It seems to be adapting well to many areas and is now almost cosmopolitan in distribution.

Green Frog

Rana clamitans Latreille

Recognition: A medium-sized brown or greenish frog with dorso-lateral folds that do not extend to the groin and with no light line along the upper jaw. Length, snout to vent, usually 2 to 3½ inches, to 4. Belly usually white to gray, with variable markings; throat dusky or yellowish, or even bright yellow in some males. Young usually with many small, dark dorsal spots and specks.

Variation: Quite variable across the state; 2 subspecies are recognized:

Rana clamitans clamitans Latreille. Bronze Frog. Usually with a plain, brown or bronzy back. Belly light, with wormlike markings; throat dusky, sometimes yellowish in males. Size smaller; length, snout to vent, usually less than 3 inches. In Kentucky, limited to the Jackson Purchase; intergrades eastward with:

Rana clamitans melanota (Rafinesque). Green Frog. Back usually green to greenish brown, with dark brown or gray-brown blotches. Belly light, barely, if at all, reticulate; throat light, that of male usually bright yellow. Size larger; length, snout to vent, often over 3½ inches. From extreme eastern Kentucky westward to near the Jackson Purchase where it meets and intergrades with the bronze frog.

Males are usually recognizable by their generally smaller size, larger tympani, stouter forelimbs, swollen thumbs, greenish upper jaws, and yellowish throats.

Confusing Species: Most like the bullfrog, but the latter has no dorso-lateral folds on the trunk.

Habitat: A variety of aquatic environments—swamps, sloughs, lakes, ponds, rivers, creeks, brooks, and springs. Roadside ditches with standing water and mudholes in little used roads often harbor young green frogs. They extend up small feeder streams, especially in wooded areas, to the head of the longest hollow, well beyond the uppermost limits of the distribution of any fish.

Kentucky Distribution: Statewide. Everywhere common, but it seems most abundant along the smaller streams in the Eastern Coal Field.

Voice: Variable, but normally an explosive, banjolike "pung." Often repeated a few times in measured succession, each note slightly lower in pitch and volume than the preceding. May be commonly heard at night and occasionally by day, especially on overcast days in May and June.

Life History: Breeding occurs from about the middle of May until mid-August. Each female deposits up to 4,000 black and white eggs in bunches which float in flattened masses on the surface of the breeding pool. The eggs are similar to the bullfrog's, but somewhat less numerous; they hatch in a matter of days, depending on a number of factors of which temperature is a major one.

The tadpoles spend about a year before metamorphosing

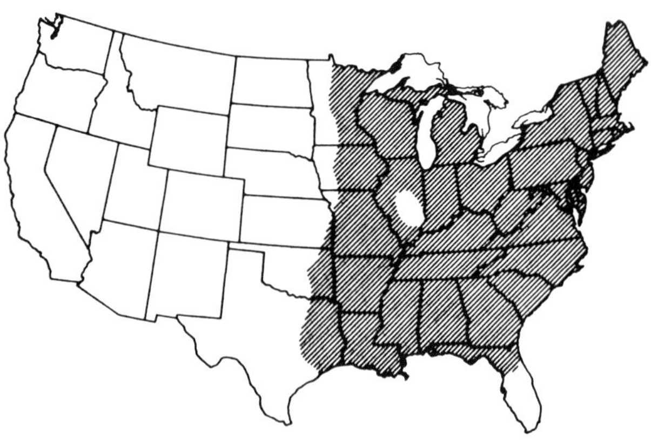

Range of the Green Frog in the United States

into frogs; metamorphosis usually occurs from April to September.

From an economic standpoint, the green frog is of little importance. Many are taken by frog hunters and eaten as small bullfrogs. Far more find their way into the stomachs of such predators as snakes, turtles, fish, mink, and raccoons, as well as larger frogs. In turn, the green frog eats any living thing it can capture, subdue, and swallow, but insects and crustaceans apparently constitute the major items in the diet.

Leopard Frog XXIV

Rana pipiens Schreber

Recognition: A small to medium-sized, green or brownish frog with 2 or 3 rows of rounded, light-edged dark spots between the complete dorso-lateral folds. Length, snout to vent, usually 2 to 3½ inches, to 4. Sides variously spotted, but belly, throat, and under surface of hind limbs white.

Variation: Highly variable in Kentucky, especially east to west. Two subspecies apparently occur in the state, but authorities do not agree on this.

Rana pipiens pipiens Schreber. Northern Leopard Frog. Usually there is a light spot in the center of the tympanum, but it is not clear white and is rather inconspicuous; sides with many dark spots. Snout short, rather blunt. Ranges from east Kentucky westward across the Blue Grass and the Knobs in the north, and to about the Mammoth Cave area in the south. Intergrades from there westward in a broad belt with:

Rana pipiens sphenocephala Cope. Southern Leopard Frog. Usually with a conspicuous white spot in the center of the tympanum; sides with few dark spots. Snout long, sharp-pointed. Occurs in the Jackson Purchase and intergrades eastward with the preceding.

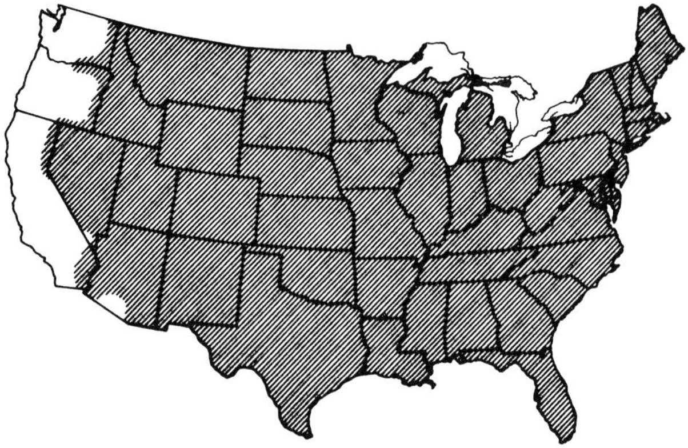

Range of the Leopard Frog in the United States

Confusing Species: Most like the pickerel frog, which has, however, 2 rows of squarish spots between the dorso-lateral folds and orange or yellow on the under surface of the hind legs.

Habitat: Bodies of water from small creeks to rivers and farm ponds to lakes. In summer it frequently wanders considerable distances from water into meadows and sometimes open woods where it finds shelter and shade beneath clumps of grass, weeds, or other vegetation. The common name "meadow frog" is understandable in such circumstances.

Kentucky Distribution: Statewide. Everywhere abundant, but it seems most common in the Jackson Purchase.

Voice: Variable, but similar to the sound created by rubbing the fingers over a tightly inflated balloon. The call has been described as a rattling snore, interspersed with assorted clucks and grunts, of one or more syllables. Males have paired vocal sacs which swell into inflated spheres above the arms when calling. It most commonly calls from March or April into May.

Life History: Emerging from the mud in March or April, the leopard frog congregates about ponds, lakes, sloughs, or other permanent, quiet water to breed.

The eggs are deposited in slightly flattened, compact spherical masses which are about 3 to 6 inches across. The egg masses are usually attached to grasses or other submerged vegetation in shallow water, but frequently lie unattached on the bottom. Each female deposits from 3,000 to 5,000 eggs per season, beginning from mid-March in the Jackson Purchase to April in the mountains.

Incubation is about a week, and the tadpoles metamorphose usually in June, July, or August.

Remarks: This is the frog most commonly dissected in various biology classes.

Pickerel Frog XXIV

Rana palustris Le Conte

Recognition: A small to medium-sized gray or tan frog with a double row of squarish blotches between the complete dorso-lateral folds. Length, snout to vent, usually 1½ to 2¾ inches, to 3⅛. Dorsal blotches often with curved margins, more nearly square than circular; sometimes 2 or more blotches coalesce, forming rectangular bars between and parallel to the dorso-lateral folds. Belly white or sometimes washed with yellow; concealed surface of hind legs yellow or orange. The dorso-lateral folds extend to the groin, and there is a light line along the upper jaw.

Variation: No geographic variation has been recorded. However, there is some evidence that the population in western Kentucky is somewhat different from that of the eastern and central areas. Males are readily recognizable by their swollen thumbs.

Confusing Species: Similar to the leopard frog, but the latter has rounded, light-edged dorsal spots and no yellow or orange on the concealed surface of the hind legs.

Habitat: Most commonly encountered about cool, clear water —lakes, ponds, or streams. In spring they gather to breed along streams and lakes, but in summer they often wander far from water and may be encountered in woods or grassy fields. They often enter caves but apparently do not regularly go much beyond the twilight zone in summer. Other frogs occasionally enter caves, but the chances are excellent that any frog encountered in a Kentucky cave is a pickerel frog. The animals ordinarily spend the winter buried in the mud, but at least a few winter well back in caves. They have been found in mid-winter several hundred feet back in a cave, well beyond the twilight zone.

Kentucky Distribution: Statewide. Although widespread, they are not commonly encountered and are rarely abundant. They are perhaps more common in eastern Kentucky than elsewhere, with a particularly high population in the cave region of Carter County.

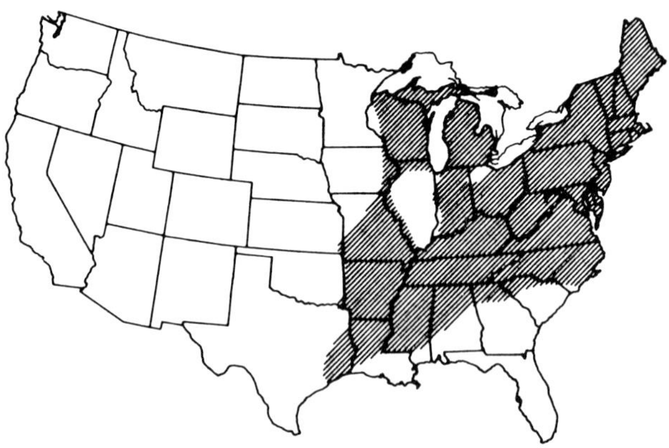

Range of the Pickerel Frog in the United States

Voice: The call is rather weak and consists basically of a grating, snorelike call with a duration of 1 to 3 seconds. Calls are uttered with equal ease from the air or under water. The male possesses a pair of expandable vocal sacs just in front of the forelimbs; air from the lungs is passed over the vocal cords, causing the call, and is stored in these sacs. When the air supply in the lungs is nearly exhausted, it is replenished from the vocal sacs and the calls repeated, using the same air over and over. Since no air is lost from the body in calling, the animal can call from under the water with no bubbling. Calls from under water are nearly as loud as those from the surface, but the calling frogs are considerably more difficult to locate.

Males usually begin calling in March and continue until May, but their calling becomes more sporadic as the season progresses.

Life History: When the frogs emerge from hibernation, usually in March, they gather about the breeding pools to mate and lay their eggs. The breeding season extends usually from late March into May, but the peak is usually reached in late March or early April.

Eggs are laid in the water, attached to submerged twigs or stems, in globular masses some 3 or 4 inches across. Each female lays from 1,000 to 3,000 eggs. They hatch in about 10 to 21 days, depending in part on temperature. After some 70 to 100 days the tadpoles metamorphose and leave the water as little frogs.

Remarks: Most of our frogs secrete an acrid, mildly poisonous fluid from the skin, and the pickerel frog does this more than most others. The animal is not dangerously poisonous, but many animals will not eat it. When it is closely confined with other species of amphibians, the others will sometimes die from the poison. Frog hunters occasionally take a few pickerel frogs, and after cleaning 2 or 3, their hands begin to burn; soap and water is an effective cure.

Wood Frog XXV

Rana sylvatica Le Conte

Recognition: A small to medium-sized, unspotted frog with a black mask on the cheek and extending through the eye. Length, snout to vent, usually 1½ to 2¾ inches, to 3¼. Dorsal color from tan or light pink through grays and browns to almost black. Although the cheek patch is usually black, it is sometimes a dark chestnut and is nearly always discernible, no matter how dark the frog. Color changes in the same individual are sometimes quite marked. Dorso-lateral folds are conspicuous and reach the groin.

Variation: No geographic variation has been observed; *Rana sylvatica sylvatica* Le Conte, the eastern wood frog, is the subspecies occurring in the state. Males can be recognized by their swollen thumbs, each with a small pad on its inner surface, their stouter forelimbs, and the convex webbing between the hind toes. In the breeding pools most females are reddish and the males darker. Both sexes are light when away from the breeding pools, but the females are generally lighter than the males.

Confusing Species: Color and pattern most like the bullfrog, which is larger, however, and lacks dorso-lateral folds and the black facial mask.

Habitat: The wood frog is usually a solitary animal, occurring in moist woodlands where there are permanent or at least semipermanent pools of water. In the summer they wander into the woods far from water and are rarely encountered. They blend so remarkably into the leaf litter of the forest floor that they are almost never seen until they leap.

Kentucky Distribution: Spotty and poorly known. Throughout the Eastern Coal Field, west to the Knobs in the north

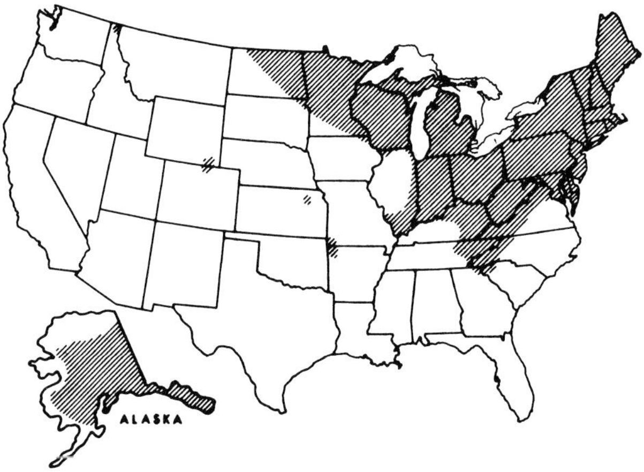

Range of the Wood Frog in the United States

and westward around the southern Knobs at least to Nelson County. Westward in the valley of the Green River to the Ohio River, with records from Adair, Edmonson, and Henderson counties. Although apparently nowhere abundant, large breeding aggregations have been found in Carter and Estill counties.

Voice: The call of the male is a series of 5 or 6 short explosive clucking notes with little carrying power. A chorus, however, may be quite loud and is strongly reminiscent of the sounds made by a flock of angry crows after an owl.

Life History: A few consecutive days and nights above freezing in early March will result in the wood frogs emerging from hibernation and assembling at the breeding pools. Usually these are permanent or semipermanent woodland pools. If not the first frogs to arrive, they are certainly a close second. I have stood half knee-deep in snow, with the temperature in the thirties, and listened to a chorus of wood frogs.

They remain at the pools no longer than 3 weeks, and each female lays from 2,000 to 3,000 eggs in sunken, globular clusters. Incubation requires only a few days. The tadpoles grow surprisingly rapidly, and are ready to metamorphose by May or June. Shortly after metamorphosis the young frogs leave the pools to take up a terrestrial existence.

Crawfish Frog XXV

Rana areolata Baird and Girard

Recognition: A medium-sized, short-legged, stubby frog with many irregular light-edged dark spots on the light-colored back. Length, snout to vent, usually 2 to 3 inches, to 3⅝. Coloration highly variable, depending on temperature and activity, but the belly is usually light and without markings. The dorsolateral folds reach the groin.

Variation: No geographic variation has been observed; *Rana areolata circulosa* Rice, the northern crawfish frog, is the only subspecies occurring in the state. Males are recognizable by their swollen thumbs and saclike vocal pouches on the sides of the head. Females are usually larger and have glandular ridges below the tympani.

Confusing Species: The numerous light-edged dark spots on a light background will readily separate this heavy-bodied species from any of our other spotted frogs.

Habitat: Terrestrial, secretive, rarely seen except in the short breeding season. Normally remains hidden in abandoned or usurped crayfish burrows, at least during the daylight hours. A large burrow, with a flattened worn spot on the edge of the entrance, is frequently occupied by a crawfish frog. It can often be seen a few inches below the entrance, but when disturbed it retreats to the bottom, sometimes 5 feet deep.

In spring they gather about farm ponds, flooded fields, or small lakes to mate and lay their eggs.

Kentucky Distribution: Known throughout the Jackson Purchase, but probably also occurs up the Ohio River valley at least to Daviess County. The species is common where it occurs, and large breeding aggregations are known in Hickman and Calloway counties.

Voice: The call is a prolonged, bass snoring sound with remarkable carrying power. A chorus is reminiscent of a pen of loudly grunting hogs. They can call from under water and in air, and an individual calling from under 18 inches of water can be heard at a distance of half a mile. Calling occurs only during the breeding season and seems to reach a crescendo on warm nights after a heavy rain during late March or early April.

Life History: Breeding occurs from early March through April, depending upon temperature and rainfall. They seem to prefer rather warm periods of heavy rainfall, when the low-lying fields are flooded and farm ponds are filled to overflowing. The eggs are laid in submerged clusters in quiet water, each female laying 5,000 to 7,000 eggs per season. The eggs hatch in a few days, and the tadpoles metamorphose in June or July.

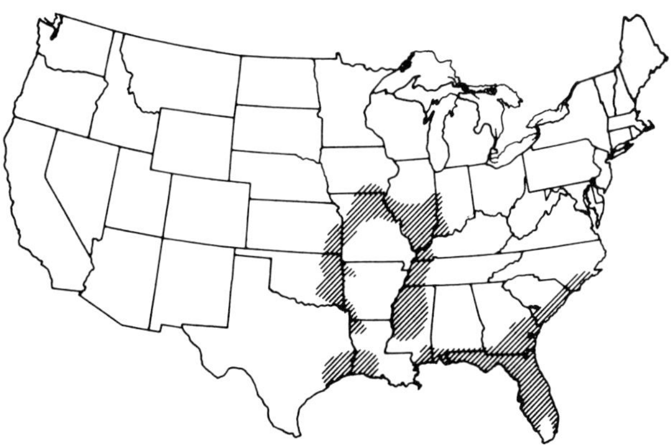

Range of the Crawfish Frog

FAMILY MICROHYLIDAE Microhylids

Apparently originating in Southeast Asia, this family has spread to New Guinea, Africa, Madagascar, and the Americas. Seven subfamilies are recognized, only one of which occurs in the United States. The largest of the subfamilies, the MICROHYLINAE, ranges from Southeast Asia northward to Manchuria and from the central United States southward to Argentina. Ten of the 16 known genera of the subfamily are Asiatic.

Two species, one with 3 subspecies, in 2 closely related genera occur in the United States. Only one species occurs in Kentucky.

Narrow-mouthed Toad XXVI

Gastrophryne carolinensis (Holbrook)

Recognition: A small, short-legged, dark toad with a tiny, pointed head, and a fold of skin across the head behind the eyes. Length, snout to vent, usually ¾ to 1¼ inches, to nearly 2. The toes are not webbed. Ground color gray, brown, or reddish, the color of a given toad varying with temperature and activity. The dorsal pattern often consists of a broad, dark, mid-dorsal area bordered on each side by lighter areas, but frequently this pattern is obscured by various blotches and mottlings. The belly is strongly mottled.

Variation: No geographic variation is recognized in Kentucky; *Gastrophryne carolinensis carolinensis* (Holbrook), the eastern narrow-mouthed toad, is the only subspecies occurring in the state. The sexes are difficult to distinguish externally, but in the breeding season the throat of the male becomes baggy and discolored.

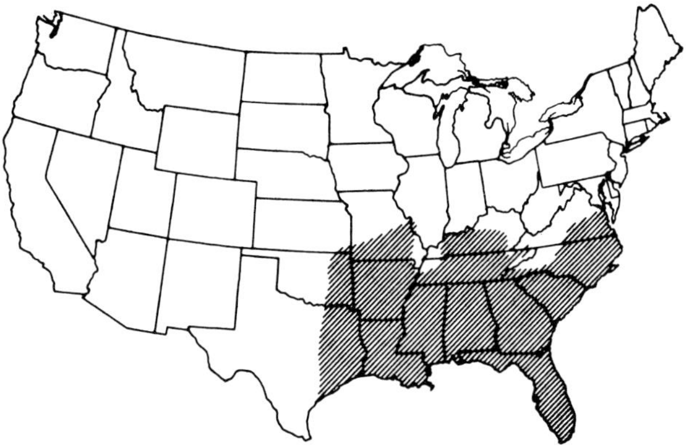

Range of the Narrow-mouthed Toad in the United States

Confusing Species: The tiny, turtlelike head and the fold of skin behind the eyes are unique among our frogs and toads; it would be difficult to confuse this animal with any other species in Kentucky.

Habitat: Shade and a moist environment are requisites for this species, but it is nearly ubiquitous. In Kentucky they have been found in such diverse habitats as beneath a rotten board in a grassy country churchyard, beneath a stone in an abandoned, brush-grown quarry, and in the shade of a buttonbush in a swampy wood. Farm ponds and swampy arms of lakes and impoundments seem favored breeding sites.

Kentucky Distribution: Poorly known. Ranges at least from Bell County westward to Marshall County; occurs northward to Madison county in the east and Meade county in the west. This species seems uncommon, but rather large breeding choruses have been recorded in Pulaski, Edmonson, and Warren counties.

Voice: The call is a low, nasal buzz, sounding remarkably like the "baa" of a lamb and rarely audible for 50 yards. A large chorus, however, can be heard at some distance. The voices of a small chorus, calling from the banks of a farm pond on a rainy, warm, spring night could easily by mistaken for the bleating of lambs.

Life History: Adults congregate about farm ponds and swampy wetlands in May and June to mate and lay their eggs; periods of heavy rainfall stimulate the breeding activities. When at the breeding pools, the males often call from burrows and crevices along the shore and quickly retreat deep into them when disturbed.

Each female lays about 850 eggs in small surface films, layer upon layer, each film containing from a few to nearly 100 eggs. The egg is unique among Kentucky amphibian eggs in that it has a flat place on it, rather than being completely spherical.

Hatching occurs in a few days, and metamorphosis occurs in about 3 to 10 weeks. The tadpole is unique in possessing a sucking disc for a mouth.

FAMILY BUFONIDAE True Toads

The Bufonidae are the toads of everyday language. The family contains some 15 genera, of which *Bufo* is the most abundant and widespread. This genus occurs on all major land masses except Greenland, Australia, New Guinea, and New Zealand.

The Bufonidae usually breed in open water, the eggs being laid in long strings on the bottom. The African live-bearing toad, *Nectophrynoides vivipara*, however, is a notable exception. Fertilization is internal, and the eggs develop in the oviducts of the female. Although the eggs produced are relatively few, as many as a hundred tiny toads have been removed from a single female. At birth the youngsters are fully formed, but they retain some larval characteristics for a time.

The skin secretions of all our true toads are mildly poisonous and are irritating to the mucous membranes. Many animals will not eat them, and there are instances of deaths in dogs from merely picking up a large toad with the jaws. None of them, incidentally, cause warts.

Bufonidae is represented in the United States and Canada by 15 species of the genus *Bufo*. In Kentucky 2 species, one with 2 subspecies, are recognized.

KEY TO THE SPECIES AND SUBSPECIES OF KENTUCKY *Bufo*

1 a. Few (usually not more than 3) warts per color spot; cranial crests conspicuous; femoral warts generally with spines (*Bufo americanus* Holbrook, American Toad) 2

b. Several (usually more than 3) warts per color spot; cranial crests rather inconspicuous; femoral warts generally without spines / *Bufo woodhousei fowleri* Hinckley, Fowler's Toad (p. 120)

1a

1b

2 a. Dorsal spots, if present, usually small and including a single wart; belly plain or sparsely spotted; size smaller; snout to vent length, less than 3 inches; head length usually more than 30 percent of snout to vent length / *Bufo americanus charlesmithi* Bragg, Dwarf American Toad (p. 118)

b. Dorsal spots larger, and often with 2 or 3 warts; belly dark spotted or mottled; size larger; snout to vent length often over 3 inches; head length usually less than 30 percent of snout to vent length / *Bufo americanus americanus* Holbrook, American Toad (p. 118)

American Toad XXVII

Bufo americanus Holbrook

Recognition: A large, warty toad with few warts in each dorsal color spot, conspicuous cranial crests, and a short spine atop most of the femoral warts. Length, snout to vent, usually 2 to 3½ inches, to 4¼. Color highly variable, from brick-red through browns and olive grays to light gray; sometimes variously patterned with light patches; and a mid-dorsal light stripe is often present. Belly spotted or not.

Variation: Extremely variable in color and pattern. Geographic variation is well marked across the state from east to west, with two subspecies represented:

Bufo americanus americanus Holbrook. American Toad. Recognizable by a combination of characters: Dorsal spots large, conspicuous, often with 2 or 3 warts in each; belly plain or sparsely spotted; size larger; snout to vent length often over 3 inches; head length usually less than 30 percent of the snout to vent length. Ranges across Kentucky from the east to the Jackson Purchase; intergrades with:

Bufo americanus charlesmithi Bragg. Dwarf American Toad. Dorsal spots often absent; if present, usually with a single wart. Belly usually dark spotted. Size smaller; snout to vent length not over 3 inches; head length usually more than 30 percent of snout to vent length. Occurs in the Jackson Purchase area; intergrades with preceding.

Males are usually smaller than females and somewhat less

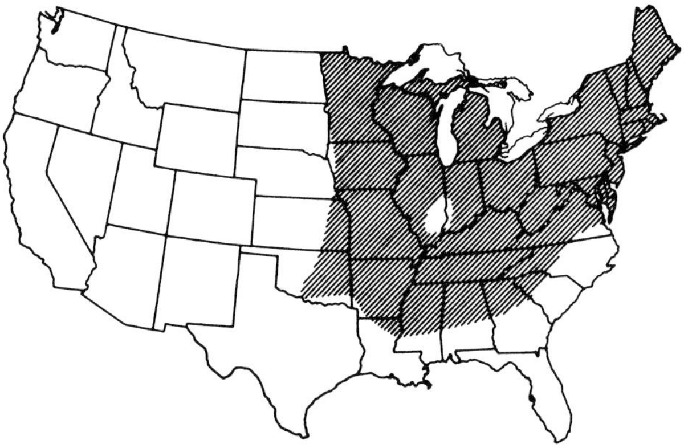

Range of the American Toad in the United States

gaudy in color. They can be distinguished by their stouter first and second fingers with horny excresences upon them and their dark, or at least dusky, swollen throats.

Confusing Species: Most like Fowler's toad, with which it sometimes hybridizes. Fowler's toad usually has more than 3 warts in each color spot, weak, inconspicuous cranial crests, and a virtually unspotted chest and belly.

Habitat: One of our most ubiquitous amphibians. Usually encountered about houses and gardens, it is often found in deep woods, far from the nearest human habitation. During the day, it usually remains secreted under a board, rock, clump of vegetation, or other hiding place, emerging at night to feed.

Kentucky Distribution: Statewide; abundant.

Voice: Beginning usually in March, the long drawn-out, high-pitched musical trills of the males may be heard from nearly every pond and roadside puddle by night. Near the peak of the breeding season, usually in mid-April, the calls can sometimes also be heard throughout the day. The voice is readily recognizable; it is easier to identify an American toad by his voice than by his appearance.

Life History: Adults congregate about the breeding pools in March or early April to mate and lay their eggs. The peak of the breeding season is generally in mid-April, 2 weeks or more before Fowler's toads begin breeding. Eggs are laid in the water in long gelatinous strings, containing as many as several thousand eggs, arranged in twisted double, or sometimes triple, rows; the strings may be several feet in length. The eggs hatch in about a week, and the resulting tadpoles are small, black, and readily recognizable as toad tadpoles. They feed and grow until late May or early June, when they metamorphose into tiny toads. The youngsters are sometimes so abundant about favored bodies of water that they almost cover the ground at the peak of their transformation.

Food of the adults consists of a great variety of invertebrate animals, mostly insects.

Fowler's Toad XXVII

Bufo woodhousei Girard

Recognition: A small to medium-sized toad with several warts in each dorsal color spot, rather inconspicuous cranial crests, and few spines on the femoral warts. Length, snout to vent, usually 2 to 3 inches, to 3¼. Dorsal coloration usually brownish or gray, occasionally greenish or brick red, with usually a light mid-dorsal stripe. Normally the belly is unspotted, but there is frequently a single dusky spot on the breast.

Variation: No geographic variation has been observed across Kentucky; *Bufo woodhousei fowleri* Hinckley, Fowler's toad, is the only subspecies occurring in the state. Males are usually smaller than females, have horny growths on the fingers, and have darker, more flaccid throats.

Confusing Species: Most like the American toad, which has 1 to 3 warts per dorsal color spot and conspicuous cranial crests. Sometimes hybridizes with the American toad and the hybrids are usually intermediate between the two; such animals are difficult to identify.

Might be confused with the spadefoot toad, but the spadefoot has no obvious paratoids, only one spur on each hind food, a smoother, less warty skin, and the pupils are vertically elliptical.

Habitat: Most commonly encountered about homes and gardens, but may be found nearly any place where there is shelter and a nearby breeding site. Usually breeds in permanent, quiet, shallow water such as farm ponds, lakes, impoundments, and drainage ditches. Roadside puddles and small streams are sometimes used.

Kentucky Distribution: Statewide; abundant. They seem least abundant in the Eastern Coal Field. In many areas in central and western Kentucky one can travel for miles on a warm May evening and never be out of hearing of a chorus of Fowler's toads.

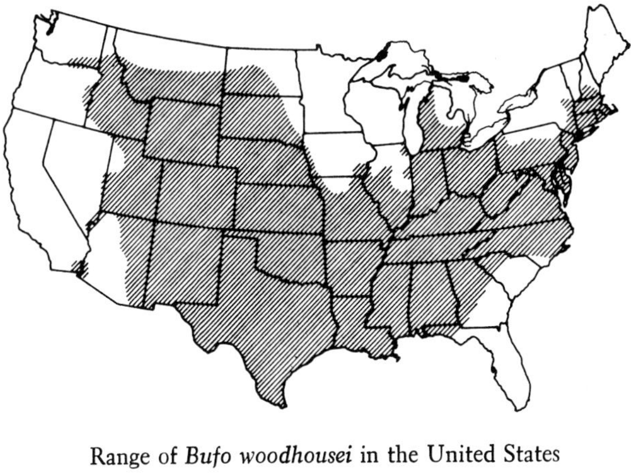

Range of *Bufo woodhousei* in the United States

Voice: The calls of the male are quite distinctive, a loud, nasal scream of 2 to 3 seconds' duration, repeated over and over. Calls are usually heard at night, but sometimes during the day, particularly when the sky is overcast. They are usually first heard in early May and may continue until July, becoming more sporadic later in the season.

Life History: In late April or early May the adults begin to congregate about the breeding pools. They reach a peak of sexual activity about mid-May and usually leave the pools by July. Eggs are laid in shallow water, in gelatinous strings as long as 10 feet or more containing as many as 8,000 eggs. Generally, the string of eggs is a twisted double strand. They may be distinguished from the eggs of the American toad by the single-walled gelatinous tube; that of the American toad is double-walled.

Hatching occurs in about a week, and the tadpoles metamorphose from late June until August. Sometimes recently metamorphosed youngsters are so abundant about the margins of farm ponds that they nearly cover the ground.

Food consists of a great variety of invertebrates, chiefly insects.

FAMILY HYLIDAE Treefrogs and Their Allies

Thirty-two genera and several hundred species of Hylidae are known, and all the genera save 2 are confined to the New World. The genus *Nyctimystes* occurs only on New Guinea, and *Hyla* is found around the world in tropical and temperate regions except in most of Africa, the southern Middle East, India, and most of the southern coast of Asia.

Most members of the family follow the usual reproductive pattern, depositing eggs in open water that hatch into tadpoles and metamorphose into miniature adults. However, there are several notable exceptions. Some species build mud basins to receive the eggs, and the tadpoles develop and metamorphose before leaving the nest. Some lay their eggs on leaves over water and the tadpoles drop into the water and develop there. *Hyla* in Jamaica lay their eggs in the water accumulating at the bases of Bromeliad leaves. Several South American genera have developed the habit of carrying the eggs about on the back of the female, exposed in some genera, covered by a flap of skin in others.

The family is almost surely tropical in origin and has radiated toward the poles. Three groups have reached Kentucky: *Acris*, the cricket frogs, represented by a single species; *Hyla*, the tree-frogs, with 4 species; and *Pseudacris*, the chorus frogs, 2 species.

Most hylids are arboreal, and exhibit rounded pads on the tips of the toes (digital discs). These are adaptations for climbing. The cricket frogs, *Acris*, have become secondarily terrestrial and have much reduced discs.

KEY TO THE GENERA AND SPECIES OF KENTUCKY HYLIDAE

1 a. With alternating light and dark vertical bars on the upper jaw; digital discs reduced; back warty; size small (less than 1⅜ inches snout to vent length); feet extensively webbed / *Acris crepitans blanchardi* Harper, Blanchard's Cricket Frog (p. 126)

1a

b. Without alternating light and dark vertical bars on the upper jaw; digital discs usually quite conspicuous; back smooth or rough, may be pebbled, but not warty; size often larger, but some species less than 1⅜ inches snout to vent length: feet slightly or not at all webbed 2

2 a. Dorsum rough, pebbled; a light spot (greenish, whitish, or yellowish) about as large as the eye on the side of the face just below the eye 3

b. Dorsum smooth; no light spot on side of face below eye 4

2a

3a

3 a. Rear of groin washed with green; bars on both fore and hind legs; the dorsal pigmented area (X or star-shaped) centering usually on rear half of the back / *Hyla avivoca avivoca* Viosca, Western Bird-voiced Treefrog (p. 134)

b. Rear of groin yellow or orange; no bars on fore legs; pigmented pattern on back centering usually on front half of back / *Hyla versicolor versicolor* Le Conte, Eastern Gray Treefrog (p. 132)

4 a. No stripes or mottling on back; a distinct white line on side, starting below the eye and continuing backward over the shoulder toward the hip / *Hyla cinerea cinerea* (Schneider), Green Treefrog (p. 130)

b. With stripes or mottling on back; no white line as above 5

4a

5a

5 a. With a mid-dorsal stripe, which usually is divided caudally / *Pseudacris triseriata feriarum* (Baird), Upland Chorus Frog (p. 136)

b. Without a mid-dorsal stripe 6

6 a. Dorsum not patterned; assorted small spots and dark flecks usually present but these not organized into a recognizable pattern / *Pseudacris triseriata feriarum* (Baird), Upland Chorus Frog (p. 136)

b. With a dorsal pattern 7

7a

7b

7 a. Dorsal pattern in the form of an oblique cross / *Hyla crucifer crucifer* Wied, Northern Spring Peeper (p. 128)

b. Dorsal pattern in the form of a pair of large, dark reversed parentheses, or a broad H; a dark triangle between the eyes extending caudally / *Pseudacris brachyphona* Cope, Mountain Chorus Frog (p. 139)

Cricket Frog

Acris crepitans Baird

Recognition: A small, warty-skinned hylid with a pattern of alternating light and dark bars on the upper jaw and a dark triangle between the eyes. The digital discs are small and poorly developed, and the feet are extensively webbed. Length, snout to vent, ¾ to 1¼ inches usually, up to about 1⅜. Color and pattern extremely variable. Color changes are pronounced, the same frog being gray, green, brown, or red, and patterned or not, depending upon temperature, moisture, light, and activity. There is a dark, ragged stripe along the concealed rear surface of the thigh, regardless of the dorsal color and pattern.

Variation: No geographic variation has been observed in Kentucky; *Acris crepitans blanchardi* Harper, Blanchard's cricket frog, is the only subspecies known in the state. Males can be recognized in the breeding season by their dark throats.

Confusing Species: Similar to *Pseudacris* (both species) and *Hyla crucifer*, but can be distinguished by the alternating light and dark bars on the upper jaw, the more warty skin, and the dark stripe on the concealed surface of the thigh.

Habitat: Seldom encountered except about the breeding area; likely it is essentially confined there. Sunny mud flats and shallow water with some vegetative cover, either emergent or along the shore, are necessary. Farm ponds, impoundments, lakes, and the larger creeks and rivers are the usual habitat.

Kentucky Distribution: Statewide; usually abundant. Although quite spotty in the Blue Grass and absent from many apparently suitable habitats, it seems least abundant in southeastern Kentucky. Extremely high populations are reached in places, particularly in the Land Between the Lakes in Lyon and Trigg counties.

Voice: From about the middle of April on through the summer, the familiar "gick, gick, gick" of the cricket frog arises from around many bodies of permanent water in Kentucky. The voice of a single male resembles the clicking of two pebbles held loosely in the hand and shaken slowly. At a distance of a few inches the call is loud and grating and soon wears on the nerves. From a distance, however, a single calling male produces a rather pleasant sound, and a chorus is delightful.

Life History: Although a hylid in structure, the cricket frog is more like a true frog (Ranidae) in habits. It does not climb well and spends its time on the ground or in shallow water. The legs are long, and it is an excellent jumper.

Cricket frogs seem reluctant to hibernate and can remain active at temperatures near freezing. They have been encountered abroad in Kentucky in every month save January. When they emerge, usually in February or March, they are already about the breeding pools, which are their normal haunts. Consequently, it seems strange that they are such late breeders. Calling begins usually about the middle or last of April, and egg laying begins shortly thereafter and continues sporadically into the summer.

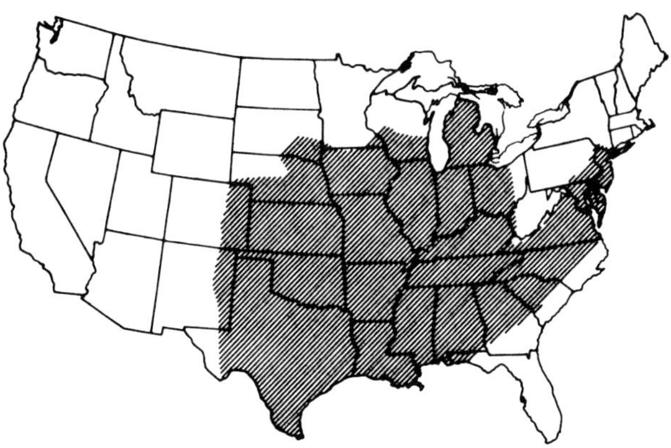

Range of the Cricket Frog in the United States

The eggs are deposited in small, floating packets, spread in a thin film and containing usually about 6 small eggs. During a season, one female may produce over 200 eggs. Hatching occurs in a few days, and the tiny tadpoles can be recognized by their black-tipped tails. Recently metamorphosed young can be encountered from late June or July into September or October, depending upon when the eggs were laid.

Food consists of a variety of tiny invertebrates, principally arthropods.

Spring Peeper XXIX

Hyla crucifer Wied

Recognition: A small hylid, bearing a more or less conspicuous X-shaped dark mark on the back. Length, snout to vent, ¾ to 1 inch usually, up to about 1⅜. Dorsal color various, through shades of gray, brown, or olive to almost black. Belly unmarked.

Variation: Spring peepers are remarkably uniform across Kentucky, *Hyla crucifer crucifer* Wied, the northern spring peeper being the only subspecies occurring in the state. Apparently no appreciable differences in voice or appearance exist. Males can be distinguished from females by their smaller size and, during the breeding season, by their darker throats.

Confusing Species: Spring peepers might be confused with three other species of tree frogs, but only the peeper has an X-shaped mark on the back. Mountain chorus frogs bear a broadly H-shaped, or reversed parentheses, mark on the back and usually have a dark triangle between the eyes. Cricket frogs are variable, but show a series of alternating light and dark spots along the upper jaw. Upland chorus frogs usually have a light line along the upper jaw and 3 straight lines down the back, the central one dividing posteriorly, or a series of spots or blotches.

Habitat: Woods and thickets are the usual habitats, but individuals are rarely encountered there. Ordinarily they are noticed only in spring when they gather in noisy breeding aggregations about shallow lakes, woodland ponds, roadside ditches, and quiet streams.

Kentucky Distribution: Every county in Kentucky apparently supports a population of spring peepers, but they now seem quite scarce in the Blue Grass. The population there has declined over the past several years, probably because of the increasing scarcity of suitable breeding pools. In most sections of the state one can drive for miles on warm spring nights and never be out of hearing of a chorus of peepers.

Voice: Peepers are often more easily identified by voice than by appearance. The usual call is a single, clear, piping whistle emitted at the rate of about one a second. There is a slight, upward slur at the end of the call, and a single note can be heard at a distance of up to half a mile. Males approaching a pond utter a peculiar plaintive, querulous call. Calling usually begins in March and continues until June. Although they mostly call by night, they often call throughout the day during the height of the breeding season.

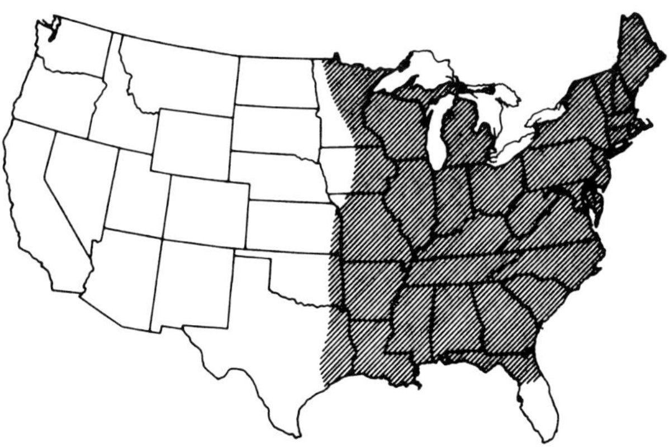

Range of the Spring Peeper in the United States

Life History: Peepers emerge from hibernation usually in early March. They gather in vast numbers about the breeding pools and may continue their breeding until June. Each female deposits up to about 1,000 eggs, which are attached singly to submerged sticks, grasses, leaves, or other objects. Hatching occurs in a few days and most tadpoles metamorphose in May or June. The offspring of the late breeders, however, may not metamorphose until July or August.

As the breeding season ends, the adults and metamorphosed young gradually leave the breeding pools and climb into the trees and bushes where they spend the summer.

Food consists of a variety of small invertebrate animals, mostly insects.

Green Treefrog XXIX

Hyla cinerea (Schneider)

Recognition: A small, plain, slender treefrog, usually bright green, with a conspicuous light stripe along each side extending from beneath the eye over the forelimb and back toward the hip. Length, snout to vent, usually 1¼ to 2 inches, to 2½. Although usually bright green, the color may be yellowish, dull greenish gray, slate-gray, or nearly black. The same individual may exhibit these colors, depending upon temperature and activity. Frequently tiny golden spots are scattered widely over the dorsum.

Variation: No geographic variation is recognized in Kentucky, and none is likely as the range in the state is so small. *Hyla cinerea cinerea* (Schneider), the green treefrog, is the only subspecies occurring in the state. Males are smaller than females and have wrinkled throats because of the vocal pouch.

Confusing Species: There is no reason to confuse this with any other Kentucky anuran. The combination of digital discs, unmarked dorsum, and lateral light stripes is distinctive.

Habitat: Seldom encountered except in breeding congresses in cypress swamps or floodplain sloughs; likely they live in or about such areas throughout the year. During the breeding season individuals are sometimes found perched on vines, tree branches, or shrubs over the water.

Kentucky Distribution: Poorly known, but it surely occurs in swamps throughout at least the western half of the Jackson Purchase. Apparently most abundant about the north end of Reelfoot Lake in Fulton County, but data on distribution and abundance are scarce.

Voice: A clear, bell-like note repeated over and over in a measured cadence, sometimes as frequently as 75 times a minute; the individual call is a nasal "quonck." In the Deep South, the animal is generally thought to be a "rain frog," whose call presages rain. The calls may be heard from mid-May into August, but are most common on warm evenings in late May and early June.

Life History: Little is known of the breeding of this species in Kentucky. Eggs are laid in small packets at or near the surface

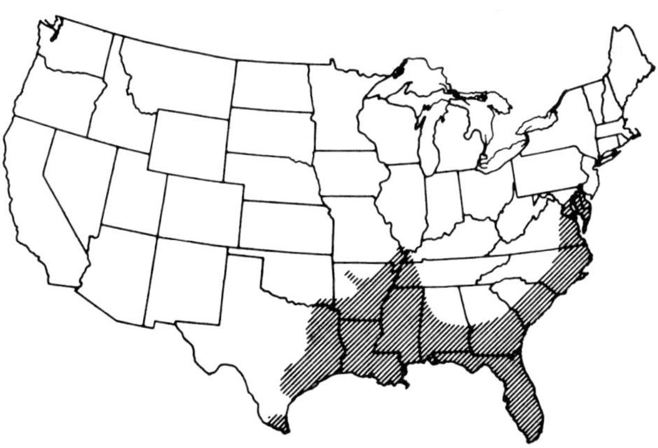

Range of the Green Treefrog in the United States

of the water. Hatching occurs in 3 or 4 days, and the tadpoles metamorphose when about 2 months old.

Remarks: This is one of the least known anurans in Kentucky; a serious study of the species in the state is sorely needed.

Gray Treefrog XXX

Hyla versicolor (Le Conte)

Recognition: A large treefrog with a dark, roughly star-shaped pattern on the back, a light spot beneath the eye, and with the concealed surfaces of the thighs washed with orange or yellow. Length, snout to vent, usually 1¼ to 2 inches, to 2⅜. Dorsal color and pattern extremely variable, even in the same individual. Dorsal color slate gray, brown, green, light gray, or almost white. Light spot below the eye white, yellowish, or greenish, but nearly always discernible, regardless of dorsal color or pattern. Back with many minute warts, rather pebbly in appearance. Belly plain, light, and unmarked.

Variation: No geographic variation is recognized in Kentucky; *Hyla versicolor versicolor* (Le Conte), the eastern gray treefrog,

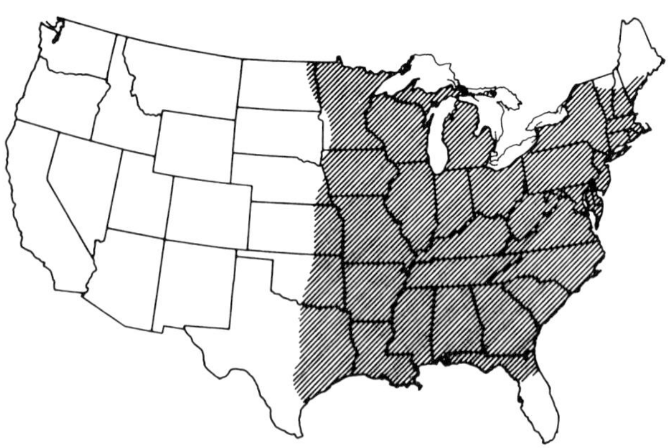

Range of the Gray Treefrog in the United States

is the only subspecies occurring in the state. Males are smaller than females and have a process on the pad at the base of the thumb that extends onto the upper part of the thumb; in the females the pad is restricted to the ventral and inner side of the thumb. In the breeding season, males have dark throats and vocal pouches.

Confusing Species: Most like the western bird-voiced treefrog, but the latter is considerably smaller with the concealed surface of the thigh washed with green and the dorsal pattern centered toward the rear of the back.

Habitat: A woodland species, restricted to trees and brush except during the breeding season, when individuals congregate about woodland pools, farm ponds, impoundments, lakes, and streams. Roadside ditches and mudholes in little used woodland roads are sometimes utilized as breeding pools. They are rarely seen on the ground except when at the breeding areas or en route.

Kentucky Distribution: Statewide; usually abundant. Scarce in the Blue Grass though in many localities the habitat appears suitable. The species is particularly abundant in Robinson Forest in Breathitt County.

Voice: The common call of the male is a loud, coarse trill lasting for 1 to 3 seconds, ending abruptly, and audible for a quarter mile or more. At the height of the breeding season, the calls are given at an average of 20 to 24 per minute. They call sporadically from April until September or October, particularly on overcast, sultry days. This is the Kentucky "rain frog" whose call supposedly foretells rain and frequently does. In late April or early May they congregate about breeding pools, and the choruses have a quality of their own, completely unlike the individual calls.

Life History: After emergence in early spring, the gray treefrogs go into the trees (about the time the leaves are well started out) where they stay and feed until late April when they begin to congregate about the breeding pools. They do not remain continuously at the pools, but move into the surrounding trees on occasion. Rainy nights bring them back to the pools time and again (or perhaps different individuals?) until August, but peak activity seems to be in late May.

Eggs are laid in packets of 2 to 3 dozen, loosely attached to vegetation at the surface of the water. They hatch in 4 or 5 days and the tadpoles transform when about 2 months old. After the breeding season they forsake the pools and go into the trees to stay until autumn. In midsummer, particularly in hot, dry weather, they take refuge in hollows in trees; perhaps some winter in such places.

Remarks: The gray treefrog is remarkably adept at changing its color to match the surroundings. It sometimes blends so well with tree bark that it is almost impossible to discern, and one sitting lengthwise on a limb looks almost exactly like a knot or a tiny snag.

Bird-voiced Treefrog XXXI

Hyla avivoca Viosca

Recognition: Like a small edition of the gray treefrog, with the dorsal pattern toward the rump and a greenish wash on the concealed surfaces of the thighs. Length, snout to vent, usually 1 to 1¾ inches, to slightly over 2. Dorsal coloration generally gray, brown, or green. A light spot is present on the side of the face below the eye, nearly always discernible, regardless of the dorsal color. Dorsal pattern as well as color highly variable, even in the same individual, depending upon temperature, environment, and activity. Belly plain, unspotted.

Variation: No geographic variation is recognized in Kentucky; *Hyla avivoca avivoca* Viosca, the western bird-voiced treefrog

Range of the Bird-voiced Treefrog

is the subspecies occurring in the state. Males are considerably smaller than females and in the breeding season exhibit a darkened throat.

Confusing Species: Much like the larger gray treefrog, in which the concealed surface of the thighs is orange or yellowish and the dorsal pattern is nearer the head.

Habitat: In Kentucky this species seems largely limited to cypress swamps. Farther south it occurs along the swampy margins of the larger creeks and rivers, as well as in the cypress, tupelo, and buttonbush tangles of the swamplands. Breeding areas are surely the same streams and swamps.

Kentucky Distribution: Poorly known, but apparently occurring in suitable habitats throughout the Jackson Purchase and up the Ohio River at least to Henderson County. A magnificent population occurs in some of the swamps of Hickman County.

Voice: A clear, ringing, whistled "wit-wit-wit-wit-wit" repeated rapidly 20 times or so. A single calling male sounds more like a bird than a frog and has fooled many a bird watcher. The males usually begin calling about mid-May and continue until August. At the peak of the breeding season, on warm humid nights in late May, the entire swampland seems to ring with the calls of thousands of individuals. Calling becomes more sporadic as the season progresses and generally ceases by August.

Life History: After emergence from winter quarters, they go into the trees, where they forage to the tops of the very tallest ones. Occasionally a lone male can be heard calling from the top of some giant cypress. With the advent of warm, humid nights in mid-May, they descend to near the surface of the water, and sit about the vines, buttonbushes, swamp roses, and small cypresses, the males calling in an ever-increasing chorus.

Essentially nothing is known of the breeding of this species in Kentucky. A single female may lay up to 600 or more eggs in a season in submerged packets. Hatching occurs in 3 or 4 days, and the tadpoles metamorphose when about a month old.

Remarks: A serious study of the biology of this animal in western Kentucky would be rewarding; any accurate observations on its biology would be worthwhile.

Chorus Frog XXXII

Pseudacris triseriata (Wied)

Recognition: A small, extremely variable hylid with a light line on the upper jaw and a dark stripe through the eye and extending along the side toward the groin. The dorsal pattern is of three basic types: 1) striped—a series of 3 longitudinal dorsal stripes, either continuous or made up of rows of spots, with the middle one dividing toward the rear; 2) spotted—assorted dorsal spots and flecks; 3) plain—dorsum without spots. Usually there is a dark triangular spot, or at least a suggestion of one, between the eyes. Length, snout to vent, usually ¾ to 1¼ inches, to 1⅜. Color variable, but usually gray or brown, with dark brown or black markings. Belly usually yellowish, sometimes with dark flecks on the breast. The feet are not webbed.

Variation: Only a single subspecies, *Pseudacris triseriata feriarum* (Baird), the upland chorus frog, is known to occur in the state. However, as one progresses northwestward across Kentucky, from Bell County in the east to McLean County in the west, there is a gradual and quite well marked decrease in the percentage of striped individuals and a corresponding increase in the plain individuals. In Rockcastle and Pulaski counties, only about 15 percent of the population is plain; in Ohio and Hancock counties, the plain individuals constitute about 55 percent of the total population. Specimens from Hickman County in the Jackson Purchase do not fit this pattern; less than 5 percent are plain. However, the striped individuals show much fainter, less obvious stripes than the more eastern population.

The percentage of spotted individuals does not correlate with geography, but varies rather irregularly, from about 30 percent to over 60 percent of the population. Casey, Clinton, and Nelson counties have the highest proportion of spotted individuals (over 60 percent) of any areas yet examined.

A male can be recognized in the breeding season by his vocal pouch; if it is not extended, then the throat is dark and flaccid.

Confusing Species: Most like the mountain chorus frog, which has a distinct H-shaped, or reversed parentheses, mark on the back. Similar to the spring peeper, which has an X-shaped dorsal pattern. Might be confused with the cricket frog, which has many warts, extensively webbed feet, and alternating light and dark bars on the upper jaw.

Habitat: Rarely seen except about the breeding pools, normally areas of shallow water with much emergent herbaceous vegetation, such as grasses and sedges. This is a creature of the grasslands and other open areas; it is rarely found in woods or even woodland edges.

Kentucky Distribution: From the counties of Bell, Knox, Laurel, Rockcastle, and southern Madison in the east westward across the state. Extends northward around the southern edge of the Knobs at least to Jefferson County. Extremely abundant throughout the range.

Voice: The common call of the male is a series of from 10 to 20 rapidly repeated clicking sounds, usually rising in pitch. It can be imitated fairly well by drawing the thumbnail across the points of a comb. Calls have been heard in Kentucky as early as February 3 and as late as December 5. This species sometimes calls when the air temperature is below freezing and ice is on the ponds. A temperature of 55° F. or higher seems to be preferred; at least, when the air temperature drops below that figure, a chorus will soon become silent.

Life History: Two or three consecutive days of continuous 55° F. or higher temperature in February or March will bring this species from its winter retreats in crayfish holes and other underground shelters. They are among our first frogs to appear and soon congregate about marshy areas and grassy roadside ditches to mate and lay their eggs.

Mating begins by late February or early March, and a peak

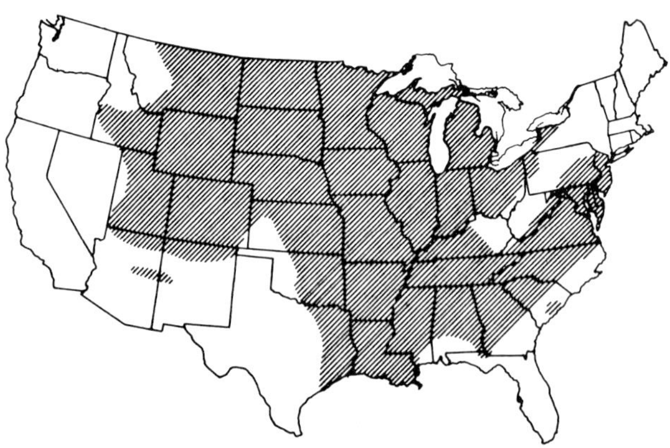

Range of the Chorus Frog in the United States

is reached about mid-March; by May sexual activity has essentially ceased for the year. Each female deposits about 100 eggs in small loose clusters attached to submerged vegetation. Incubation is but a few days, and the tadpoles metamorphose in May or June, when they are about 60 days old.

Mountain Chorus Frog XXXII

Pseudacris brachyphona Cope

Recognition: A small hylid with a pair of large dark marks, like reversed parentheses, on the back. There is also a dark triangular spot between the eyes and a light line along the upper jaw. In some specimens the dark dorsal marks curve so far inward as to touch in the middle of the back, forming a broad, irregular X- or H-shaped mark. Length, snout to vent, usually 1 to 1¼ inches, to 1½. Color varies from gray through brown to nearly black. Dorsal markings dark, brownish or black. Ventral surface light, unmarked; concealed surface of hind limbs yellowish.

Variation: No geographic variation is recognized. Males are somewhat smaller than females and in the breeding season may be recognized by their dark throats and vocal pouches. If the pouch is collapsed, it leaves the throat flaccid.

Confusing Species: Most like the spring peeper, which has a rather well defined narrow X-shaped mark on the dorsum and no light line on the upper jaw. Similar to some individuals of the upland chorus frog, but they never have dorsal marks shaped like reversed parentheses. Might be confused with the cricket frog, which has alternating light and dark bars on the upper jaw, or with a small woodfrog, which has extensively webbed hind feet, conspicuous dorso-lateral folds, and no obvious dorsal markings.

Habitat: A creature of the woodlands, emerging in early spring to congregate about springs, spring runs, small woodland streams, pools, and roadside ditches in the woods or at woodland edges. Sometimes puddles in little-used roads serve as breeding pools. It is rarely encountered along streams large enough to support game fish.

Kentucky Distribution: Throughout the Eastern Coal Field, extending westward in the Green River drainage at least to Edmonson County and northward through the Rolling Fork drainage into Jefferson County. It reaches its highest population in the mountain counties, where one can travel miles on a warm early spring evening and never once be out of the sound of their voices.

Voice: I first became familiar with this creature when I was a small boy in one of the mountain counties of Kentucky, where it was called a "reke-rake." Perhaps this colloquial name is still applied, but I haven't heard it for many years. The name is derived from the voices of several males calling in a chorus. The individual call is often a short, rasping "rake" repeated at short intervals. The pitch varies considerably from one animal to the next (perhaps older ones are deeper?), and some utter a sound more nearly like "reke." The frogs nearly always call in

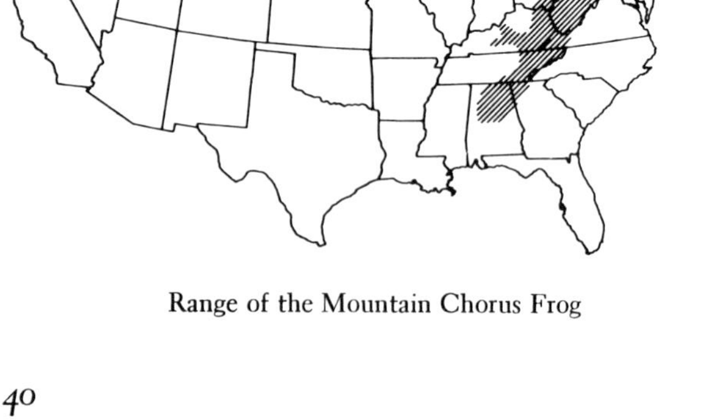

Range of the Mountain Chorus Frog

a chorus, and more likely than not, adjacent animals will be differently pitched and will alternate calls. From a few feet away, such a duet sounds like one animal "reke-rake, reke-rake, reke-rake . . ."—hence the name.

Calling begins as early as February 12, but reaches a peak in late March; by late April or early May the frogs are essentially silent. However, a day or so of rain in summer will sometimes bring them back to the breeding pools to call as late as August.

Life History: Mountain chorus frogs ordinarily come out of hibernation and go to the breeding pools in late February or early March, depending on the weather. Generally, they can be found calling at the breeding pools after the first three consecutive nights in February or March when the temperature does not go below 50° F. Once they arrive at the pools, they generally neither call nor breed when the temperature drops below 50° F.

Females lay about 1,000 to 1,100 eggs each breeding season in small clusters of from 25 to 50 eggs. These clusters are attached to twigs, leaves, or other debris at or near the bottom of shallow puddles or other small, quiet pools. Under usual weather conditions the eggs hatch in about a week or 10 days. Under abnormally warm conditions, however, they will hatch in 3 days. The tiny tadpoles feed and grow for about 30 to 45 days and then metamorphose into miniature adults. These tiny animals leave the breeding pools and go into the woods. Many do not survive, but those that do return to the puddles the following year as breeding adults.

By late April or early May the adults have usually left the breeding pools and gone back into the woods, where they may sometimes wander considerable distances. After a day or so of rain, however, they will frequently return to the breeding puddles for a few days to breed again, even into June or July. Breeding aggregations and freshly laid eggs have been found after several days of rain as late as August on Big Black Mountain in Harlan County.

LIST OF PLATES (AMPHIBIANS)

(Above) Hellbender, *Cryptobranchus alleganiensis alleganiensis*

(Below) Western Lesser Siren, *Siren intermedia nettingi*

II

(Above) Jefferson's Salamander, *Ambystoma jeffersonianum*

(Below) Triploid Jefferson's Salamander, *Ambystoma platineum*

(Above) Small-mouthed Salamander, *Ambystoma texanum*

(Below) Mole Salamander, *Ambystoma talpoideum*

IV

Marbled Salamander, *Ambystoma opacum*

(Above) Spotted Salamander, *Ambystoma maculatum*

(Below) Eastern Tiger Salamander, *Ambystoma trigrinum tigrinum*

VI

Red-spotted Newt, *Notophthalmus viridescens viridescens* (aquatic phase)

Red-spotted Newt, *Notophthalmus viridescens viridescens* (terrestrial phase)

VIII

Three-toed Amphiuma, *Amphiuma means tridactylum*

Northern Dusky Salamander, *Desmognathus fuscus fuscus*

Black Mountain Dusky Salamander, *Desmognathus welteri*

X

(Above) Allegheny Mountain Salamander, *Desmognathus ochrophaeus ochrophaeus*

(Below) Allegheny Mountain Salamander, *Desmognathus ochrophaeus ochrophaeus* (old male)

Appalachian Seal Salamander, *Desmognathus monticola monticola*

XII

(Above) Eastern Red-backed Salamander, *Plethodon cinereus cinereus*

(Below) Zig-zag Salamander, *Plethodon dorsalis dorsalis*

(Above) Ravine Salamander, *Plethodon richmondi richmondi*

(Below) Slimy Salamander, *Plethodon glutinosus glutinosus*

XIV

Four-toed Salamander, *Hemidactylium scutatum*

(Above) Kentucky Spring Salamander, *Gyrinophilus porphyriticus duryi*

(Below) Northern Red Salamander, *Pseudotriton ruber ruber*

XVI

Midland Mud Salamander, *Pseudotriton montanus diastictus*

Green Salamander, *Aneides aeneus*

XVIII

Two-lined Salamander, *Eurycea bislineata*

(Above) Long-tailed Salamander, *Eurycea longicauda longicauda*

(Below) Three-lined Salamander, *Eurycea longicauda guttolineata*

XX

Cave Salamander, *Eurycea lucifuga*

XXI

Mudpuppy, *Necturus maculosus maculosus*

Eastern Spadefoot, *Scaphiopus holbrooki holbrooki*

(Above) Bullfrog, *Rana catesbeiana*

(Below) Bronze Frog, *Rana clamitans clamitans*

(Above) Northern Leopard Frog, *Rana pipiens pipiens*

(Below) Pickerel Frog, *Rana palustris*

(Above) Eastern Wood Frog, *Rana sylvatica sylvatica*

(Below) Northern Crawfish Frog, *Rana areolata circulosa*

XXVI

Eastern Narrow-mouthed Toad, *Gastrophryne carolinensis*

(Above) American Toad, *Bufo americanus americanus*

(Below) Fowler's Toad, *Bufo woodhousei fowleri*

XXVIII

Blanchard's Cricket Frog, *Acris crepitans blanchardi*

(Above) Northern Spring Peeper, *Hyla crucifer crucifer*

(Below) Green Treefrog, *Hyla cinerea cinerea*

Eastern Gray Treefrog, *Hyla versicolor versicolor*

Western Bird-voiced Treefrog, *Hyla avivoca avivoca*

XXXII

(Above) Upland Chorus Frog, *Pseudacris triseriata feriarum*

(Below) Mountain Chorus Frog, *Pseudacris brachyphona*

Reptiles

REPTILE TOPOGRAPHY

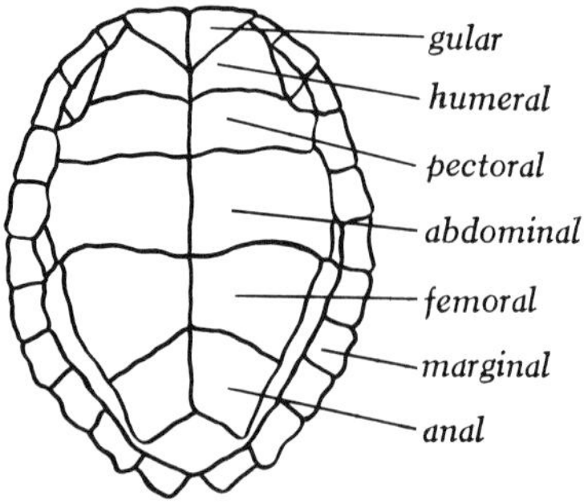

TURTLE PLASTRON

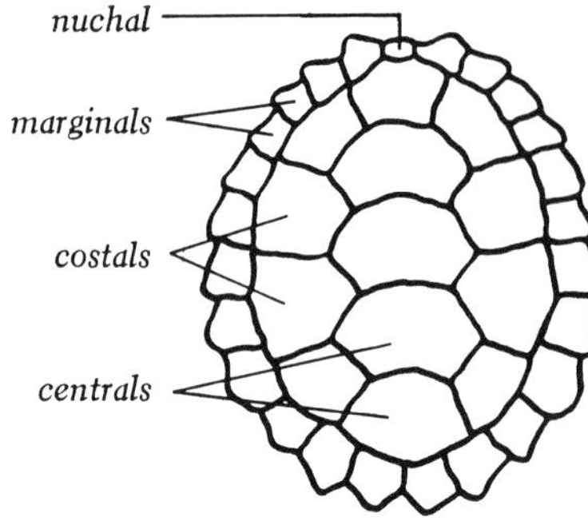

TURTLE CARAPACE

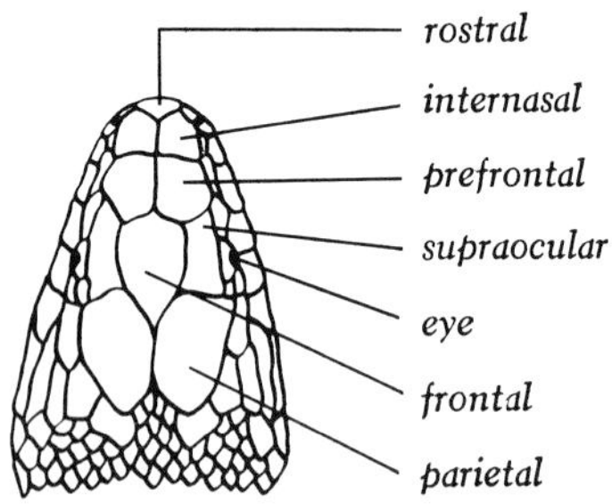

SNAKE HEAD (DORSAL VIEW)

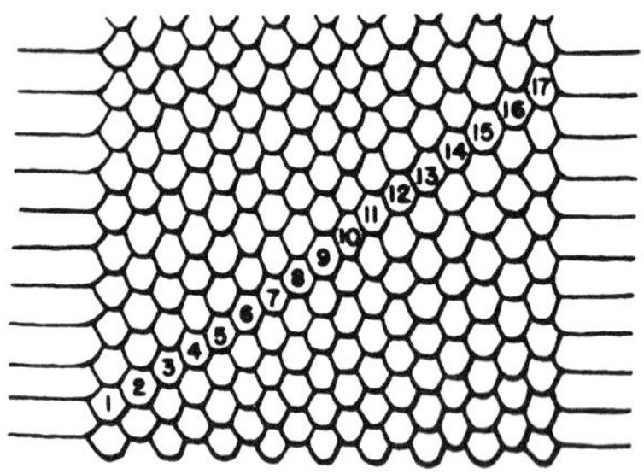

NUMBERING SNAKE SCALE ROWS

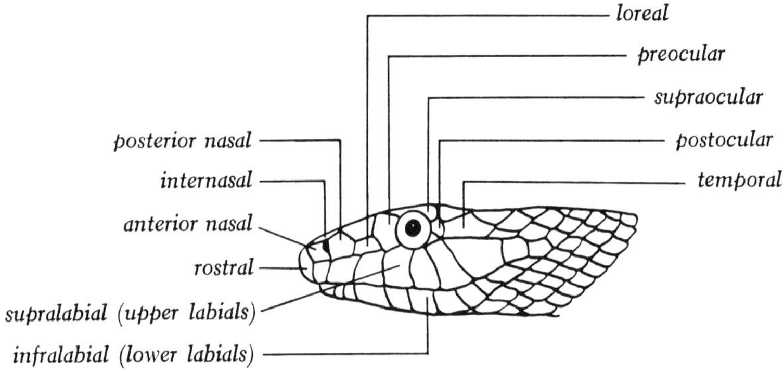

SNAKE HEAD (LATERAL VIEW)

REPTILIA

The story of the snake in the Garden of Eden is but one example of the fear and loathing with which those creatures and other reptiles have been regarded by most human beings. A great store of folklore has arisen about them, which, although in part based upon truth, is riddled with twisted facts, misrepresentations, guesses, and outright fabrications.

It is quite true that some reptiles have the ability to do great bodily harm to man. These species, however, are seldom encountered and generally attempt to escape when they are found. Only 4 of the 55 species of Kentucky Reptilia are venomous, and the average person's chances of being hurt by a reptile are infinitely smaller than his chances of being crushed and broken by an automobile.

The body temperature of reptiles, like that of the amphibians, fluctuates with the environment. Such animals depend upon an external supply of heat; consequently they are called ectothermic (=outside heat). Under certain circumstances, reptiles are able to maintain a body temperature a few degrees above the surrounding temperature, but not much and not for long. This characteristic, of course, affects the distribution of reptiles on the earth; they are most abundant in areas where the temperature does not fall below freezing. Those extending into freezing areas must be able to find nonfreezing spots to spend the winter.

Perhaps the most remarkable advance exhibited by the reptiles over the amphibians was the development of a shelled egg complete with an embryonic membrane, the amnion. The combination of shell and membrane completely freed them from the necessity of laying their eggs in water or moist places. With the advent of the shelled egg the need for a larval period ended; development was direct, from egg to miniature adult. The reptiles thus were the first vertebrates to successfully colonize the great land masses of the earth. The amphibians, the first terrestrial vertebrates, were in many ways irrevocably tied to a moist environment. By a string of fortuitous circumstances, the reptiles were freed from such limiting ties. The shelled egg, with the amnion, was certainly a major step, but other developments also contributed: 1) a scaly epidermis, which effectively cuts down cutaneous water loss; 2) well-developed lungs, which enable reptiles to acquire an ample oxygen supply in dry areas. Other minor developments came along: claws appeared on the toes; skin glands were almost completely lost; the development of a palate, separating the oral and nasal passages was started, and to this day the crocodilians are the only reptiles to have completed it.

Although the original development of reptilian characteristics admirably fitted their possessors for a terrestrial existence, several groups have become secondarily adapted to an aquatic existence; however, they still retain those characteristics that initially fitted them so well for the land. All the aquatic species still have lungs, most have scales, and nearly every one of them comes ashore to lay its shelled eggs.

About the time the amphibians were in their heyday (some 300 million years ago) there began to appear in the fossil record some transitional forms between them and reptiles, and beginning by late Triassic and early Jurassic times and continuing on through the Cretaceous period (roughly a span of 100 million years, starting about 175 million years ago) reptiles were the ascendant animals on earth. They assumed many bizarre shapes, and some became the largest animals ever to walk the face of the earth, before or since. Sixteen orders are known to have existed then, consisting of about 170 families.

There is little doubt that the reptiles arose from an amphibian stock, and it is interesting, at least, to speculate on their spectacular abundance and equally spectacular decline. Obviously, the oncoming reptiles were suited for a terrestrial existence, being quite independent of a moist environment. They encountered little effective competition on land; there were no birds and no mammals. Consequently, all habitats were open to them; nearly any kind of bizarre creature could, and in fact many did, find a place where it could survive. Such ideal conditions for speciation rarely become available, on this planet, at least.

Factors accounting for the decline of the reptiles are much less clear, but two major developments about the beginning of their decline may well have been contributing factors, if in fact not the causes.

Many of the reptiles had become so huge that they could not sustain themselves on land and were compelled to live in water to support their massive bodies. The drying of the great swamps and the formation of massive mountain ranges some 75 to 150 million years ago must surely have eliminated many of these wondrous beasts. It was about the time of these geologic changes that some effective competition appeared in the form of mammals. Obviously these primitive little mammals could not harm a monstrous dinosaur directly, but they could well have taken a fearful toll by destroying the eggs.

Modern reptiles survive in a wide variety of habitats. Some are arboreal, some terrestrial, some burrowing, some live in fresh water, and some are marine. Even the aquatic ones are tied to the land; nearly all come ashore to deposit their eggs or bear their young. Some few of the sea snakes, however, have become completely marine; their young are born at sea and never leave it.

Unlike amphibians, reproduction in reptiles shows little originality. Fertilization is internal; the sperm is introduced into the cloaca of the female by a special structure. In snakes and lizards the copulatory organ lies embedded in the base of the tail, resulting in a slight swelling that can often be recognized. The organ is paired, one on either side, and each part is separate, complete, and sometimes exceedingly spiny. They are everted from the cloaca by turning inside out, and the one inserted into the cloaca of the female is determined by which side is next to her when copulation occurs. The existence of these hemipenes is directly responsible for the folktale that snakes really have 2 hind legs, normally hidden in the body, that can be protruded to aid in locomotion.

Some reptiles lay eggs, some retain the eggs in the body until they hatch, then bear living young, and others are viviparous. In these latter the young develop within the body of the mother and obtain their food and oxygen supply directly from her via a membrance in which the circulatory systems of the mother and young interdigitate.

Reptiles as a group constitute a minor part of the vertebrates inhabiting the world today. There are only about 6,000 species, arranged in some 45 families. These families in turn are organized in 4 orders:

1) Order Testudinata. Turtles.
2) Order Rhynchocephalia. Represented by a single species, the Tuatara (*Sphenodon punctatus*), found only in New Zealand and on nearby islands.
3) Order Squamata.
    Suborder Sauria—Lizards
    Suborder Serpentes—Snakes
4) Order Crocodilia. Crocodiles, alligators, and their allies. Two living families, world-wide in tropical and subtropical regions.

KEY TO THE ORDERS OF KENTUCKY REPTILIA

1 a. Body rounded, incased in a bony box, covered with large epidermal plates or a leathery shell; jaws horny, toothless; legs present / Order TESTUDINATA, Turtles (p. 150)

b. Body elongate, covered with small scales; jaws with teeth; legs present or absent 2

2 a. Eyelids and external ear openings present; usually with legs; if legless, then a longitudinal groove on each side of the body lined with minute scales / Order SQUAMATA: Suborder SAURIA, Lizards (p. 195)

b. Eyelids and external ear openings absent; legless / Order SQUAMATA: Suborder SERPENTES, Snakes (p. 222)

1a

2a

2b

ORDER TESTUDINATA Turtles

Everyone recognizes a turtle, of whatever size, shape, color, or origin. Would that the terms applied to them were so straightforward—"turtle," "tortoise," "terrapin"—they mean one thing to one person, something else to another. Biologically all members of the order Testudinata are correctly called turtles. Terrapins and tortoises are certainly turtles, regardless of whatever else they may be. The term *tortoise* is usually applied to terrestrial turtles, particularly the larger ones; terrapin is usually applied to edible, more or less aquatic hardshelled turtles. Local usage varies, however. In Kentucky, the hardshelled, terrestrial, inedible *Terrapene* is often called a "terrapin." The "tortoise shell" of commerce comes not from tortoises, but from the marine hawksbill turtle.

Turtles, as such, are of ancient lineage. About 200 million years ago one of the early groups of stem reptiles accomplished some rather surprising evolutionary feats, few of which are preserved in the early known fossil record. This group developed a horny, toothless beak and a hard, bony box about their shortened, flattened bodies. Perhaps the most remarkable feat was getting the girdles by which the limbs are attached to the body inside the rib cage. This unlikely series of events resulted in the establishment on the face of the earth of one of the most conservative animals ever seen—a turtle.

It is not clear just when turtles did appear, but the fossil record indicates that perfectly typical turtles were here, in numbers, as long as 190 million years ago. They saw the rise and fall of the dinosaurs, the origins of the birds, and had been in existence for at least 120 million years when our oldest recognizable contemporary mammal, the opossum, appeared a mere 70 million years ago.

Today, turtles seem to be diverging in 3 major directions, all of which tend toward a reduction in the shell. Some turtles are becoming more terrestrial, with high-domed, thinner shells (weight reduction) and stronger legs and feet for support. Others are becoming more nearly aquatic, bottom feeding animals, with flattened carapaces, reduced plastrons, and more webbing on the feet and legs. The free-swimming, marine turtles constitute the third trend, toward a pelagic existence. The carapace and plastron are often reduced, and the limbs modified into flippers. The fore flippers are much enlarged and move up and down, much like the wings of a bird; these turtles literally fly through the water.

In reproduction, turtles are quite conservative. Fertilization is internal, by copulation, and all are oviparous. The eggs are laid on land, nearly always buried by the female. When the clutch is complete, she departs and leaves them to their fate.

Turtles are probably the most economically important of the reptiles. They are eaten everywhere they occur. Not only do they supply the gourmet with green turtle soup, but they and their eggs furnish a ready supply of protein in many places of the world where protein is hard to come by.

There are only about 335 species of turtles, arranged in some 65 genera, in the world today. These are grouped into 10 families, 3 of which occur in Kentucky.

KEY TO THE FAMILIES OF KENTUCKY TESTUDINATA

1 a. Shell rigid, not leathery and pliable; nose not tubular 2

b. Shell soft and leathery, not bony and rigid; nose tubular / Family TRIONYCHIDAE, Softshell turtles (p. 189)

1b

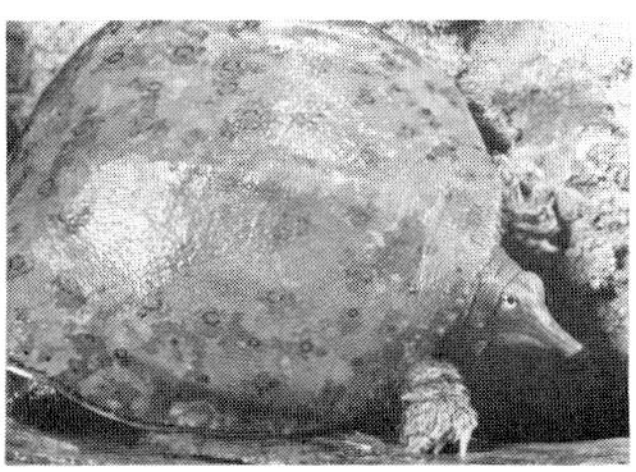

2 a. Plastron of 11 or fewer plates, cruciform or broad / Family Chelydridae, Snappers, musk and mud turtles (p. 152)

b. Plastron of 12 plates, broad / Family Emydidae, Fresh water and marsh turtles (p. 166)

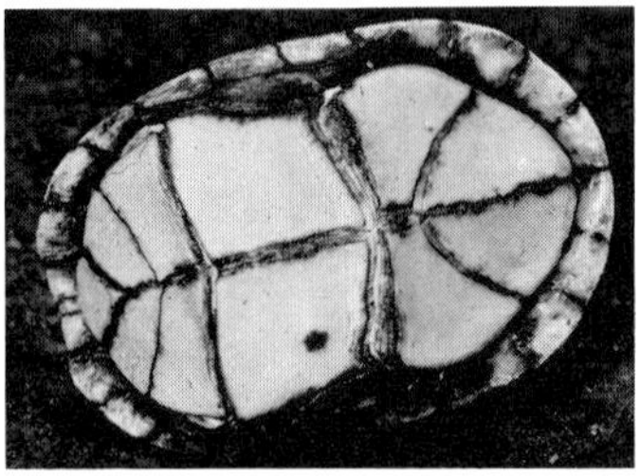

2a(1)

2a(2)

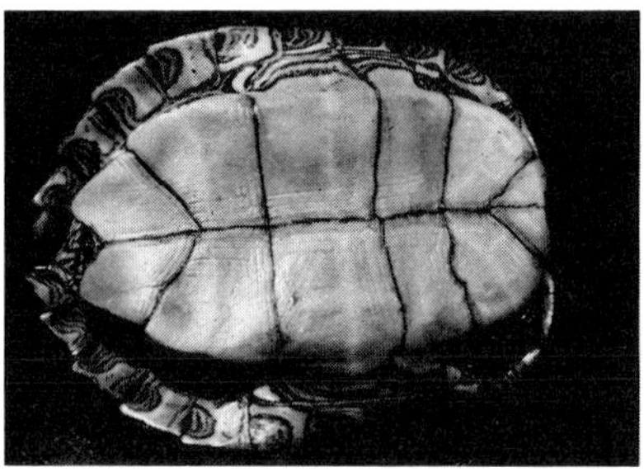

2b

FAMILY CHELYDRIDAE Snappers, Musk and Mud Turtles

This small family is restricted to the New World. It includes both our largest and smallest fresh-water turtles, the alligator snapper, *Macroclemys temmincki*, and the diminutive mud turtle, *Kinosternon subrubrum*. Three subfamilies are recognized, each with 2 living genera. One is confined to Central

America, but the other 2 range from Canada to South America. The Chelydrinae (snapping turtles) and Kinosterninae (musk and mud turtles) both occur commonly in the United States; some authors consider these separate families.

The Chelydrinae are large turtles, with large, ugly heads, small, poorly developed plastrons, long tails, and evil tempers. The Kinosterninae are small but have relatively large heads. They have better developed plastrons (particularly the mud turtles), short tails, and variable tempers—some vicious, some not.

Although essentially carnivorous or scavengers, these turtles eat considerable quantities of vegetable matter. Most are fond of basking, but many bask only while floating or lying in shallow water with the back exposed. Some, however, frequently emerge from the water to bask in the full rays of the sun.

Four species, each in a different genus, are recognized in Kentucky.

KEY TO THE GENERA AND SPECIES OF KENTUCKY CHELYDRIDAE

1 a. Tail nearly as long as carapace and bearing longitudinal keels; plastron narrow, cross-shaped, and poorly developed 2

b. Tail short, without keels; plastron larger, better developed, and not cross-shaped 3

1a

2 a. Carapace of adult essentially smooth, without prominent keels; top of head covered with smooth skin; no row of scutes between marginals and costals / *Chelydra serpentina serpentina* (Linnaeus), Common Snapping Turtle (p. 155)

b. Carapace with prominent longitudinal keels; top of head covered with smooth plates; an extra row of scutes between the marginals and the costals / *Macroclemys temmincki* (Troost), Alligator Snapping Turtle (p. 158)

2a

2b

3 a. Plastron with but one poorly-developed transverse hinge; pectoral scutes rectangular / *Sternotherus odoratus* (Latreille), Stinkpot (p. 160)

b. Plastron with two well-developed transverse hinges; pectoral scutes triangular (*Kinosternon subrubrum*) 4

3a

3b

4a

4 a. With two well-developed light stripes on each side of head / *Kinosternon subrubrum hippocrepis* (Gray), Mississippi Mud Turtle (p. 163)

b. With spots or mottling on sides of head; no obvious stripes / *Kinosternon subrubrum subrubrum* (Lacépède), Eastern Mud Turtle (p. 163)

Snapping Turtle XXXIII

Chelydra serpentina (Linnaeus)

Recognition: A snapping turtle with a large head covered with smooth skin, a long tail, and a small plastron. There is a series of saw-toothed keels on the upper side of the tail. Length usually 6 to 12 inches, to 18½. Weight to over 70 pounds. The brown to almost black carapace bears 3 rows of fairly well defined longitudinal keels, which become less conspicuous with age. Tail nearly as long as the carapace. Carapace often covered with algae. The young look quite like the adults, but are nearly black; the keels on the carapace are more evident, and the tail is longer than the carapace.

Variation: No geographic variation is recognized in Kentucky; *Chelydra serpentina serpentina* (Linnaeus), the common snapping turtle is the only subspecies occurring in the state. Males are smaller than females, and have the anal opening nearer the tip of the tail but it is difficult to distinguish the sexes externally.

Confusing Species: Quite similar to the alligator snapping turtle, which has an extra row of scales between the marginals and the costals, well developed dorsal keels, and large scales covering the head.

Habitat: Aquatic, but often makes substantial overland journeys. Any body of water in Kentucky capable of supporting game fish is likely to support a population of snapping turtles. They seem to prefer larger, quiet waters, but occur in small, clear streams.

Kentucky Distribution: Statewide. The tremendous increase in the number of farm ponds in the last few years has resulted in a considerable increase in the number of snapping turtles. They are perhaps least abundant in the Eastern Coal Field but occur in fair numbers in suitable streams throughout even that area.

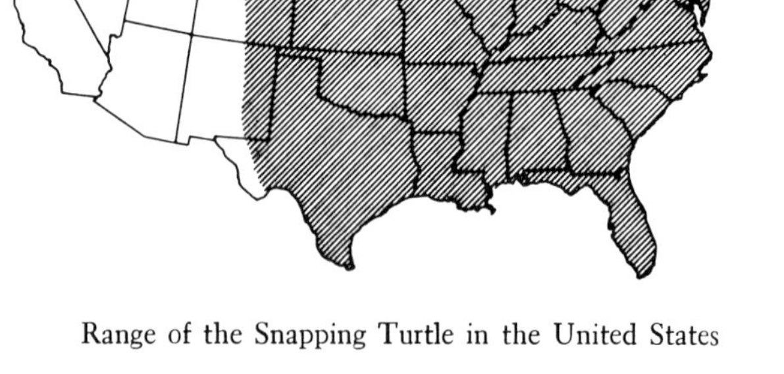

Range of the Snapping Turtle in the United States

Life History: Snapping turtles spend the winter buried in the mud under water. They favor muskrat holes and hollows under sunken logs, but have been found in many other places. Usually by October or November they are hidden away for the winter and do not regularly reappear until about mid-April. They must arouse occasionally during the winter, for they have been seen moving slowly across the bottom under a thick layer of surface ice.

Egg laying occurs usually in June. The female leaves the water early in the morning to seek out a spot to deposit her eggs, sometimes going half a mile from the water before finding an acceptable site. A flask-shaped cavity, sometimes a foot wide at the bottom, but not more than a few inches deep, is dug by alternate strokes of the hind feet. When the cavity is dug to her satisfaction, she deposits up to about 40 eggs in it, at a rate of about one a minute. Each egg as laid is directed to the bottom of the nest by a hind foot. When the clutch is complete, she fills in the cavity, stroking the loose earth over the eggs with the hind feet until little, or no sign of the nest remains.

The egg is spherical, a little over an inch in diameter, with a leathery, white shell. When dropped on a hard surface it more frequently bounces than breaks.

The eggs hatch from late August to October. At hatching the youngsters have a carapace length of from 1 to 1¼ inches and soon dig their way out of the earth to begin their travels to the water. How they find their way remains a mystery.

The snapping turtle is almost completely carnivorous, but sometimes does eat vegetable matter. Prey is about equally divided between vertebrates and invertebrates. Although carrion is sometimes eaten, most of the prey is secured by stalking it. On occasion, however, a snapper will lie quietly on the bottom and ambush his prey. Crayfish and snails are the most commonly eaten invertebrates, but a great variety of other species are taken as the opportunity presents itself. Fish and frogs are undoubtedly the commonest vertebrates taken, but salamanders, lizards, snakes, mammals, and even birds are captured and devoured. In some places, and at certain times, untold numbers of young ducks and other aquatic birds are pulled under by snapping turtles to be devoured at leisure.

Remarks: Snapping turtles are particularly aggressive when on land. Lifting their rear high in the air, they strike almost faster than the eye can follow. If they hit, they hang on doggedly. If they miss, they lose their balance and tumble forward; righting themselves, they seem enraged and will advance, carrying the attack to the enemy. In the water, they are docile and rarely make an effort to bite.

With care, a snapper on land can be captured and carried safely by the tail. One need not worry much about the hand, but watch the leg. The turtle can, and will bite a leg as it is swung by—always carry the animal with the plastron toward your leg.

The common snapper is surely the most economically important of our turtles. Thousands are eaten yearly; they are delicious when well cooked, and a delightful soup can be made from them.

Alligator Snapping Turtle XXXIII

Macroclemys temmincki (Troost)

Recognition: A large snapping turtle with a huge head covered with symmetrical plates, a hooked beak, prominent dorsal keels, and a row of extra scales on each side of the carapace between the costals and the marginals. Carapace length usually 12 to 20 inches, to over 26. Weight to 250 pounds. Color dull brown to black, with a series of 3 well-marked longitudinal keels on the carapace of the adult as well as the young. Carapace often covered with algae. The young are similar to adults, but darker, the longitudinal keels are more pronounced, the tail is longer, and the body is more papillose.

Variation: No geographic variation is recognized.

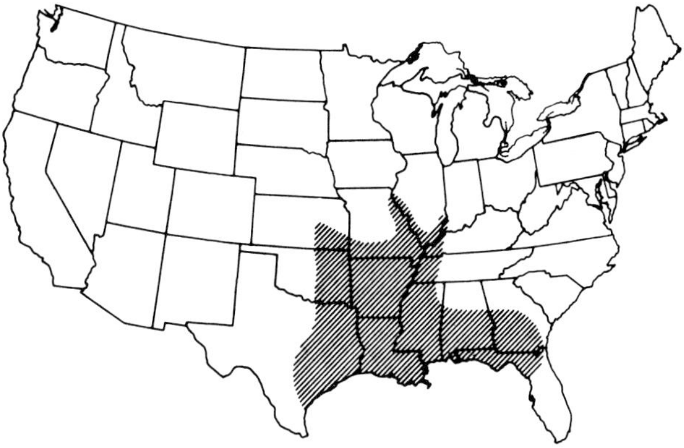

Range of the Alligator Snapping Turtle

Confusing Species: Similar to the common snapping turtle, but the latter has the top of the head covered by smooth skin rather than plates, has no extra scales on the sides between the costals and the marginals, and has less conspicuous longitudinal keels on the carapace.

Habitat: Large rivers and lakes, sometimes swamps, bayous, and larger ponds. It seems to prefer muddy bottoms with dark retreats in the form of holes, logs, or sheltering vegetation.

Kentucky Distribution: Western Kentucky, in the Mississippi, Ohio, Tennessee, and Cumberland rivers. Probably occurs up the Ohio at least to Henderson County, and perhaps into the lower Green River. Likely occurs in most larger streams, swamps, and lakes in the Jackson Purchase. This species is nowhere common, but fisherman occasionally catch one.

Life History: Little is known of the life history of this rare animal in Kentucky. The eggs are spherical, nearly 1½ inches in diameter, and have a white, leathery shell. Fifteen or 20 eggs constitute a clutch.

One of the more interesting features of this beast is its habit of "angling." On the upper surface of the tongue is an elongated, rodlike, pink structure, quite muscular, and attached by its middle to the surface of the dark-colored tongue. Both ends of the little wormlike rod are free and the turtle can wiggle them at will.

When "angling," the turtle half buries itself in the mud, opens its massive jaws, and exposes the little pink "worm" on the tongue. Wiggling it, the snapper lures fish within reach. Apparently fish constitute the major portion of the diet, but other animals are eaten as well.

Remarks: This species seems not so aggressive or evil-tempered as the common snapping turtle, but is not to be taken lightly. A 103-pound turtle in Illinois easily walked about with a 165-pound man standing on its back. A large specimen can readily break a broomstick with his jaws.

Stinkpot XXXIV

Sternotherus odoratus (Latreille)

Recognition: A small, aquatic turtle with an oval, high-domed carapace, and a large head usually with 2 yellowish stripes on each side. The plastron is relatively small and has a single inconspicuous and poorly developed hinge, situated between the 2nd and 3rd pairs of scutes; however, the hinge is not developed in the young. The pectoral scutes are roughly quadrangular, and there are barbels on both the chin and the throat. Carapace length usually 3 to 4½ inches, to 5⅜. Carapace smooth, varying in color from olive-brown to almost black and sometimes streaked or spotted with dark markings. In old individuals the head stripes become less apparent, and the carapace becomes more elongate. Young are black, usually with some light markings on the marginal scutes; the plastron is marbled with black and gray or cream. The carapace has a prominent mid-dorsal keel.

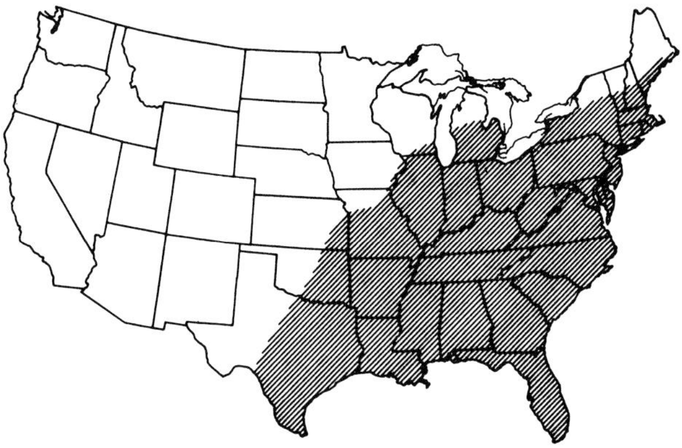

Range of the Stinkpot in the United States

Variation: No geographic variation is recognized. An adult male differs from a female in having a longer tail, a prominent claw on the tail tip, and 2 small horny patches on the hind leg that meet when the knee is bent.

Confusing Species: Most like the mud turtle, which has triangular pectoral scutes. The young are sometimes confused with the snapping turtle, which, however, has a long, keeled tail.

Habitat: Thoroughly aquatic except during the breeding season; found in permanent bodies of water, from creeks and rivers to swamps, ponds, and lakes, but prefers shallow, mud-bottomed ponds and lakes.

Kentucky Distribution: Statewide. Abundant in western Kentucky, but less so in the Eastern Coal Field. It occurs in sluggish streams in the mountains, but is absent from the higher elevations.

Life History: There is some evidence that stinkpots hibernate in large groups buried in the mud bottoms of ponds or streams, but they often hibernate singly. They usually emerge about the middle of April and congregate in shallow water where mating occurs. Apparently some mating occurs in the autumn, but the principal season of sexual activity seems to be in April and early May. By mid-May most males have left the shallows. Ovulation occurs in late May, but the eggs remain in the oviducts for about a month while albumen and shell are being secreted.

The eggs are elliptical and vary considerably in size and shape, averaging about ¾ by 1 inch. They are surrounded by a thick, brittle shell, remarkably resistant to desiccation.

During the last of June the females leave the shallows to lay their eggs on land near the water. Laying occurs at almost any hour. As soon as the female finishes laying she returns to the water. There seems little uniformity in the sort of site selected—under logs, under assorted debris, in holes dug by the hind feet, in cow tracks, in cavities dug in muskrat houses; some of the eggs are even deposited singly on the surface and not covered at all. Surprisingly enough, many of these uncovered eggs can be found to contain viable embryos.

Although as many as 9 eggs have been found in the oviducts of a female, the usual clutch contains from 3 to 5 eggs. Hatching apparently occurs in from 60 to 90 days, depending on temperature, and the youngsters make their way unassisted to the water.

Food is perhaps most noteworthy for its variety. Although chiefly carnivorous, stinkpots eat considerable quantities of vegetable matter. They will devour almost any small aquatic creature they can catch, and many terrestrial ones that are unfortunate enough to fall into the water. Occasionally they feed upon carrion.

Remarks: The stinkpot exhibits two protective actions. Handled recklessly, it will bite viciously. The neck is long, enabling the jaws to reach an unwary finger anywhere on the front half of the carapace. When disturbed a stinkpot emits a yellowish fluid from two glands opening along the edge of the carapace on either side. One is near where the rear of the bridge meets the carapace, and the other about half way between the front of the bridge and the front of the turtle. The fluid has a strong, musky odor that most likely is defensive; certainly it is responsible for the technical as well as the common name of the species.

Mud Turtle XXXIV

Kinosternon subrubrum (Lacépède)

Recognition: A small, flattened, oval turtle whose back edge is nearly perpendicular. The head is large and irregularly streaked, spotted, or marbled with light colors. The plastron is relatively large and has two obvious hinges, one between the 2nd and 3rd pairs of scutes, and the other between the 3rd and 4th pairs. However, the hinges are not developed in the young. The pectoral scutes are triangular. Carapace length usually 3 to 4 inches, to nearly 5. Carapace smooth and brownish, varying from light to almost black. Youngsters have a very rough carapace that is black, or very dark brown. The plastron is sometimes black, but often is darker at the center and along the juncture of the scutes. The lighter areas are yellowish through orange to red; each marginal scale bears a light spot. This is a handsomely marked baby turtle.

Variation: Highly variable across the state from east to west. Two subspecies are recognized:

Kinosternon subrubrum subrubrum (Lacépède). Eastern Mud Turtle. The head is variously spotted or mottled, but not usually conspicuously striped. Ranges from west-central Kentucky westward into the Jackson Purchase where it meets and intergrades with:

Kinosternon subrubrum hippocrepis (Gray). Mississippi Mud Turtle. Head usually with 2 light stripes on either side. Occurs in swamps and streams along the Mississippi River, and intergrades with the preceding where the ranges meet.

A male of this species can be distinguished from a female by his longer tail, which bears a more prominent claw at the tip, and the 2 small horny patches on the inner surface of each hind leg.

Confusing Species: Most like the stinkpot, which has a single poorly developed hinge on the plastron and quadrangular pectoral scutes.

Habitat: Basically aquatic, but less so than the closely related stinkpot. Prefers shallow water such as ponds, swamps, washes, boggy meadows, and small streams. Occasionally encountered some distance from open water in soggy places.

Kentucky Distribution: Poorly known, due in part to the secretive nature of the animal. Occurs throughout the Jackson Purchase and Land Between the Lakes, and eastward. Eastern limits in the state unknown but probably to about a line between Meade and Clinton counties. Nowhere abundant, though fairly common in the Jackson Purchase, particularly in Hickman County.

Life History: This species spends the winter buried in the mud below frost line. Some, at least, select an area out of the water to hibernate, but in wet areas where mud is but a few inches down.

Little data are available on emergence in the spring. Mud turtles are often abroad in early April and likely emerge on warm days in late March. Mating occurs in April or May, and egg laying starts soon thereafter. The peak of egg laying apparently occurs in June but is known to extend from March to September in different parts of the range.

When ready to lay, a female leaves the water and selects a rather dry spot to construct a nest and deposit her eggs. She digs a slanting hole with her hind feet. When the cavity is completed to her satisfaction, she deposits from 2 to 5 eggs in it (average about 3) and packs dirt tightly over them with her hind feet, merely covering the eggs. When she departs, the area is generally quite recognizable as a turtle nest. Ordinarily there is a semicircular cavity, about 3 x 5 inches, slanting at an angle of about 30°. The eggs are out of sight, buried in the bottom of the hole. After a few rains, all evidence of the nest is usually obliterated.

The eggs are oval and quite variable in size, but average about ⅝ by 1 inch; they are white, with a thick, hard, brittle shell.

The hatchlings are tiny, about an inch long, and ½ inch or so wide. They may hatch in September to emerge and make their way to the water, but at least some of them must surely spend the winter in the nest or the eggs overwinter; youngsters have been plowed out of nests in April, before the eggs for the current year were laid.

Food is varied, consisting mostly of invertebrates and small vertebrates.

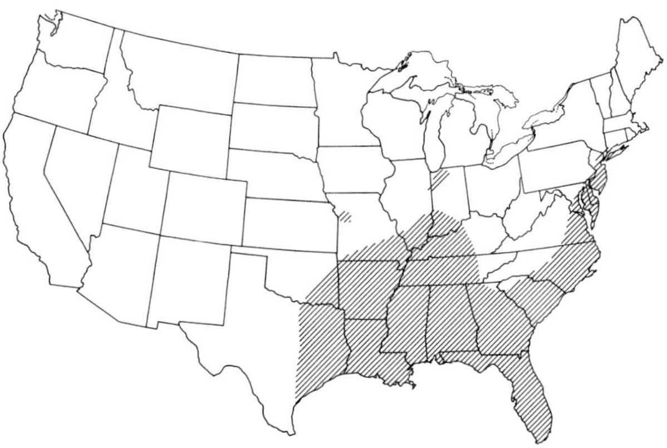

Range of the Mud Turtle

FAMILY EMYDIDAE Freshwater and Marsh Turtles

This large group of turtles is the one really flourishing family; it occurs on all continents save Australia. Some authors lump them with the true land tortoises, including the giant lumbering beasts of the Galapagos and Seychelles islands, into the family Testudinidae.

Most of the species are aquatic or semiaquatic, but some, for example the box terrapins, are adapted to an almost completely terrestrial existence.

Over half the species (8 of 14) and probably well over 75 percent of all individual turtles in Kentucky are members of this family.

KEY TO THE GENERA AND SPECIES OF KENTUCKY EMYDIDAE

1 a. Terrestrial; plastron large, hinged, and capable of closing completely / *Terrapene carolina carolina* (Linnaeus), Eastern Box Turtle (p. 170)

b. Aquatic; plastron not hinged or capable of closing 2

1a

2a, 3a, 4b

2 a. Carapace serrated on posterior edge; marginals without red pattern; no red stripes on legs or tail 3

b. Carapace not serrated on posterior edge; red patterns on marginals; red stripes on legs and tail *(Chrysemys picta)* 9

3 a. Dorsal keel present; carapace brown with yellow markings *(Graptemys)* 4

b. Dorsal keel absent; carapace greenish 6

4a

4 a. Keel on carapace with slight knobs; a yellow triangular spot behind the eye; no plastral pattern / *Graptemys geographica* (Le Sueur), Map Turtle (p. 172)

b. Keel on carapace well developed with prominent knobs; a yellow line or crescent behind eye, but no triangular spot 5

5a

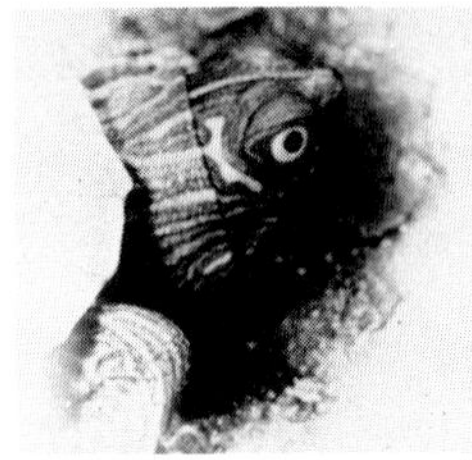

5b

5 a. Some neck stripes reaching eye, and a yellow spot or line behind the eye; no plastral pattern / *Graptemys pseudogeographica ouachitensis* Cagle, False Map Turtle (p. 175)

b. Neck stripes not reaching eye; a yellow crescent behind eye; plastral pattern well developed / *Graptemys kohni* (Bauer), Mississippi Map Turtle (p. 178)

6a

6b

6 a. Lower jaw rounded ventrally; a large red or yellow stripe behind the eye (*Chrysemys scripta*) 7

b. Lower jaw flattened ventrally; no red or yellow stripe behind eye 8

7 a. A red stripe behind eye; dark blotches on plastron, underside of marginals, and on bridge / *Chrysemys scripta elegans* (Wied), Red-eared Turtle (p. 183)

b. A yellow stripe behind eye; plastron with ocellated

pattern; underside of marginals and bridge not heavily blotched with dark / *Chrysemys scripta troosti* (Holbrook), Cumberland Turtle (p. 183)

8a

8 a. Plastron with dark markings; a dark C-shaped mark on the second lateral scale of the carapace / *Chrysemys concinna hieroglyphica* (Holbrook), Slider (p. 186)

b. Plastron without dark markings; no C-shaped light mark on the second lateral scale of the carapace / *Chrysemys floridana hoyi* (Agassiz), Missouri Slider (p. 187)

9 a. Dorsal stripe wide, prominent; plastral pattern usually absent; if present, then small / *Chrysemys picta dorsalis* (Agassiz), Southern Painted Turtle (p. 179)

b. Dorsal stripe narrow, or absent; plastral pattern well developed / *Chrysemys picta marginata* (Agassiz), Midland Painted Turtle (p. 179)

9a

9b

Box Turtle

Terrapene carolina (Linnaeus)

Recognition: A land turtle with a high, domelike shell and a broad plastron. The plastron is hinged, and both front and rear sections are movable, enabling the animal to completely inclose the body in a bony box. The fit is so precise that in many individuals not even a thin knife blade can be inserted between the carapace and plastron when the turtle is tightly closed. Carapace length usually 4 to 6 inches, to 6½.

Color and pattern are extremely variable on both carapace and plastron, often olive, yellow, orange, or even red on a black or brown ground color. Either the light or the dark colors may predominate.

Youngsters have a strongly keeled carapace that is about 1¼ inches long, and the markings are less abundant than on the adults. Often the plastron has a central dark spot and yellow margins. The hinge is nonfunctional in the young.

Variation: The only subspecies occurring in Kentucky is *Terrapene carolina carolina* (Linnaeus), the eastern box turtle. However, there is not only extreme variation in the color and pattern within a given geographic area, but there is some variation across the state that is correlated with geography. In the Jackson Purchase, one sometimes encounters box turtles that have only 3 toes on each hind foot, showing evidence of intergradation with a more western subspecies that in part is so distinguished. However, only a small percentage of the specimens from the Purchase are 3-toed, and the population seems clearly the eastern subspecies.

Males can usually be recognized by their red eyes; those of females are usually brown. In males, the posterior half of the plastron is usually concave; it is flat, or even convex, in the females.

Confusing Species: The complete closure of the shell exhibited by this species is unique among our turtles.

Habitat: Woodlands, from low swampy woods to high, dry, upland ridges. In dry, hot, weather, box turtles often congregate in small woodland pools and streams, sitting in shallow water or dug down into the mud in shady spots.

Kentucky Distribution: Statewide. Least common in the Blue Grass and the more open sections of the Mississippian Plateau, otherwise abundant.

Life History: Winters are spent hibernating, in the earth beneath the frost line. Hibernation ordinarily begins in October or November, but individuals have been seen abroad as late as December 10 in Letcher County. They emerge in April usually and mate soon thereafter. In May or June the female deposits from 2 to 7, but usually 4 or 5, eggs in a cavity in the ground, often in an open, sandy area slightly higher than the surrounding terrain; the ridge between the tracks of a logging road is a favorite site. Usually she begins digging in the late afternoon. She plants her front feet firmly and keeps them in place as she excavates the flask-shaped cavity with her hind feet. She digs, scrapes, and palms out the dirt, using the hind legs alternately and depositing the loose earth behind her.

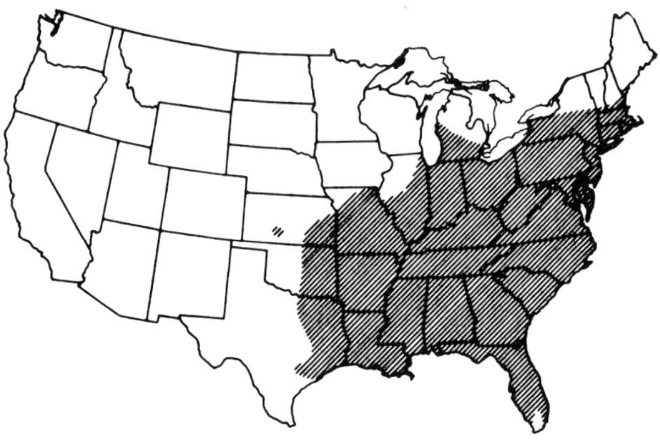

Range of the Box Turtle

When the cavity is as deep as she can reach (usually about 3 inches), she deposits her eggs in it. She then sweeps the loose earth back into the hole with her hind feet, tramping it firmly in place, layer upon layer, until no sign of the disturbance is left. Then she departs, not to return.

The box terrapin is an omnivorous feeder, and its diet changes seasonally as well as with increasing age. Youngsters are largely carnivorous, feeding mostly on slugs, insects, worms, small snakes, and other animals. Adults are more nearly herbivorous, feeding to a large extent on fungi, fruits, and berries. On occasion, a terrapin takes up residence in a vegetable garden and feeds on ripe tomatoes. They often feed on poisonous mushrooms, and seem completely immune. However, the poison remains in the terrapin, and may well be fatal to a person who eats such a terrapin.

Remarks: Box terrapins have no important predators. Forest fires kill thousands, but happily this menace is becoming of less importance. Dogs (particularly hounds) often ineffectually gnaw them. Many are killed on the highways by witless or curiously sadistic drivers who deliberately run over them.

Map Turtle XXXV

Graptemys geographica (Le Sueur)

Recognition: A small to medium-sized brownish turtle with an intricate series of light markings on the carapace that resembles a map of twisting roads; rear edge of carapace is serrate. There is a low mid-dorsal keel, but there are no or at most only low knobs arising from it. There is an isolated yellow spot behind the eye. Carapace length: males usually 4 to 6 inches, to 6½; females 7 to 10 inches usually, up to 11.

Ground color of the carapace is olive, tan, or brown. The head is dark, and the markings are yellowish. Plastron usually plain, or with various faint markings.

Hatchlings have a strongly keeled carapace with strong spines, but these become less prominent with increasing age. The carapacial pattern of the youngsters is conspicuously maplike, with many fine lines on a greenish ground color. With increasing age the pattern becomes less distinct, and the ground color changes to olive or brown.

Variation: No geographic variation is recognized. Males are considerably smaller than females and have a longer, more narrow head, a longer tail, and the anal opening closer to the tail tip.

Confusing Species: Most like the Mississippi and false map turtles, but in each of these the keel is knobbed in the adult as well as in youngsters. Additionally, the Mississippi map turtle has a crescent behind the eye, and some of the neck stripes reach the eye in the false map turtle.

Habitat: Larger rivers and lakes, rarely in ponds or small, swiftly flowing streams. It is often found in quiet backwaters and sloughs. Aquatic, this species rarely wanders on land except during the egg laying season. It is, however, a confirmed basker, often sunning itself on stranded logs or grassy or sandy areas at the water's edge.

Kentucky Distribution: Poorly known. Apparently ranges from a line roughly connecting Boyd and Bell counties westward to the Jackson Purchase. It is not clear whether the presumed absence of this species from extreme eastern and western Kentucky is due to a real absence of the turtle or to a real gap in our knowledge of the animal.

Life History: Map turtles hibernate in mud at the bottom of shallow lagoons and other quiet water. They begin emerging in early April and are out basking commonly on warm days in early May. Mating occurs soon after they emerge, and by late May or early June the females are ready to lay their eggs. Egg laying reaches a peak by the middle of June but continues into August.

The females leave the water in search of a suitable nest site, usually selecting a spot where the earth is loose, such as a sandy area or a freshly plowed field, but sometimes digging the nest in heavy clay. Occasionally they wander as far as a quarter mile from the water seeking a suitable site.

A flask-shaped cavity is dug with the hind feet, usually to a depth which the foot can conveniently reach. When the burrow is complete, the female deposits from 10 to 16 eggs in 2 layers in the expanded portion of the burrow, with the last 2 or 3 eggs in the neck, about 2 inches below the surface of the ground. After laying the eggs, the female rakes the loose earth back into the cavity and then crawls back and forth over the site a number of times, concealing the nest site.

The eggs are elliptical, dull white in color, and about 1¼ x 13/16 inches. The shell is soft and easily indented.

Hatchlings are abroad by July, but it seems that eggs laid in June hatch in August or early September. Sometimes the hatchlings from a nest do not emerge until the following May or June. Whether they overwinter as eggs or baby turtles, overwinter they surely must, for there is no other reasonable explanation for their appearance at such an early date.

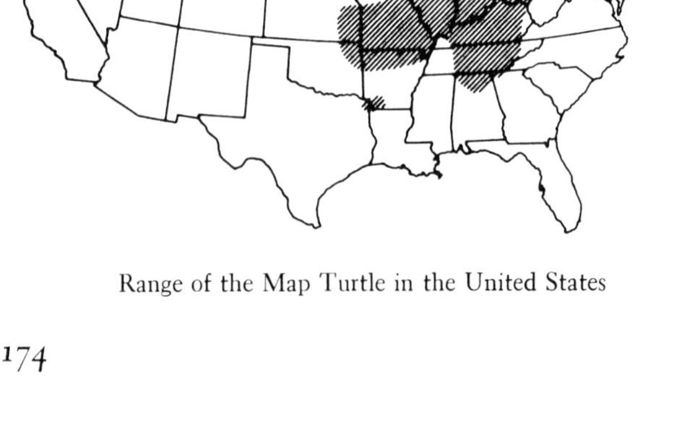

Range of the Map Turtle in the United States

Map turtles apparently feed almost exclusively on molluscs, although crayfish and insect remains have been found in their stomachs. The broad crushing surface of the jaws is admirably suited to the task of crushing snails and clams.

Remarks: This is a particularly wary turtle, diving into the water from its basking perch and swimming rapidly away at the slightest disturbance. Since they are mollusc feeders, it is difficult to bait them to a trap. Consequently, they are rarely collected, and our knowledge of them, particularly of their distribution, is poor.

False Map Turtle XXXVI

Graptemys pseudogeographica (Gray)

Recognition: A small to medium-sized brownish turtle with various light markings on the carapace and with strong serrations on the rear of the carapace. There is a strong mid-dorsal keel with dark knobs rising from it. Some of the neck stripes reach the eye, and there is a yellow spot or line behind the eye. Carapace length: males, usually 3 to 4 inches, to 4½; females, usually 5 to 9 inches, to 10.

Ground color of the carapace is usually brown, and the head is dark with yellowish markings. Plastron has a well-developed pattern.

Juveniles have a more nearly circular carapace with a conspicuous keel and high spines. The ground color is greenish, with numerous concentric figures. The plastron contains a large dark central blotch. With increasing age, the keels, spines, and pattern become less conspicuous, and the ground color changes to olive, tan, or brownish.

Variation: No geographic variation is recognized in Kentucky; *Graptemys pseudogeographica ouachitensis* Cagle, the Ouachita false map turtle, is the only subspecies known to occur in the state. Males are substantially smaller than females, have long claws on the fore feet, a longer tail, and a more posteriorly located anus.

Confusing Species: Most like the map and Mississippi map turtles. The map turtle has a low keel with no knobs or spines; the Mississippi map turtle has a crescent behind the eye.

Habitat: This highly aquatic turtle inhabits larger rivers and lakes, often in sloughs and backwaters. It appears to shun ponds, small streams, and clear streams with gravelly or sandy bottoms, preferring mud or mucky bottoms with abundant aquatic vegetation. It is a confirmed basker, often emerging from the water to lie in the sun on stranded logs or sandy or grassy areas at the edge of the water.

Kentucky Distribution: Throughout western Kentucky. Eastern limits of distribution uncertain but it is apparently unknown north or east of the Green River drainage. There is a possibility that this species occurs in some of the larger streams that empty into the Ohio in northern and eastern Kentucky.

Life History: Although sometimes seen crawling on the bottom under ice in mid-winter, this species usually spends at least the coldest parts of the winter buried in the mud bottom of lakes and rivers or buried in the underwater entrance to a muskrat den.

They emerge usually in April and by May may be seen basking in the warm spring sunshine. They are especially wary and prefer to bask on stranded logs well away from shore. The slightest disturbance will send them tumbling into the water to swim away and secrete themselves in submerged vegetation.

Little is known of the breeding habits of this species in Kentucky. Apparently they lay their eggs late in the season, in July or August. The females leave the water and select a loose plot of soil for the nest cavity. Digging is done with alternate sweeping movements of the hind feet, and a flask-shaped cavity some 5 inches deep, with a 2-inch neck and a rounded cavity 3 inches in diameter at the bottom, is soon dug. The cavity slopes forward under the body at an angle of about 60°, and the loose earth is pushed up the slanting incline with the soles of the hind feet.

When the cavity is complete, the female deposits from 7 to 13, but usually 9 or 10 eggs in the cavity and fills it in with the loose earth before returning to the water.

No data are available on incubation time, or growth and development of the young.

This species is much more vegetarian in its diet than the map turtle. There is evidence that the young are carnivorous or omnivorous, but it seems that adults are largely vegetarians. However, they sometimes take insects or feed on dead fish.

Remarks: Although wary and hard to catch, this species is highly regarded as food in many sections of the Mississippi Basin; it is generally considered the best of the "terrapins" in the area.

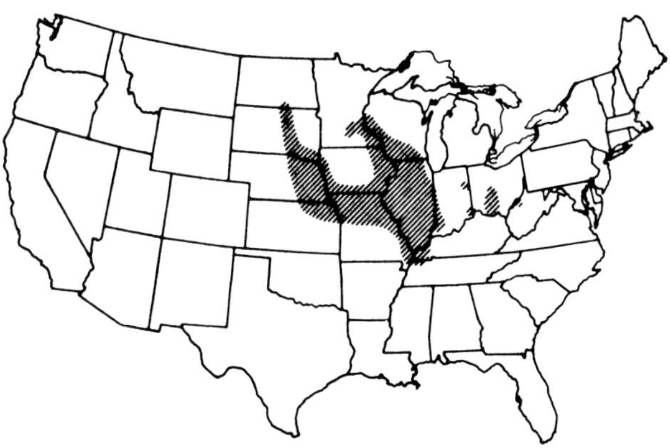

Range of the False Map Turtle

Mississippi Map Turtle

Graptemys kohni (Bauer)

Recognition: A small to medium-sized brownish turtle with an intricate series of light markings on the carapace and with serrations on the rear of the carapace. There is a prominent mid-dorsal keel, with prominent dark knobs or spines arising from it. A long, narrow, crescent-shaped spot behind each eye effectively prevents the neck stripes from reaching the eye. Carapace length: males usually 3½ to 5 inches, to 5½; females usually 6 to 9½ inches, to 10.

The carapace is olive, tan, or brown, with light markings. The head is brown or black, with yellowish lines. The plastron is greenish yellow with a variable dark pattern, usually a broad set of concentric dark lines with a lighter center.

Young are more rounded and have a more obvious keel and spines than adults. They have brighter, more contrasting colors and patterns, but these become subdued with increasing age and disappear in some adults.

Variation: No geographic variation has been observed.

Males are considerably smaller than females and have greatly elongated toenails on the forefeet.

Confusing Species: Most like the map and false map turtles. The false map turtle has the keel and knobs, but lacks the crescent behind the eye, and one or more of the light neck lines reaches the eye. The map turtle lacks the knobs, and the keel is much reduced; a yellow spot lies behind each eye.

Habitat: Quiet sloughs and backwaters of the larger rivers, impoundments, and lakes, especially where there is much aquatic vegetation and muddy bottoms.

Kentucky Distribution: Known only from the Tennessee River. Almost surely occurs throughout the Jackson Purchase and the Land Between the Lakes in suitable habitats.

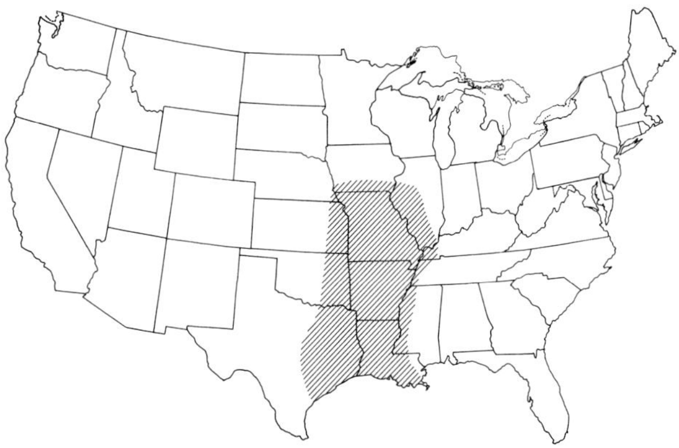

Range of the Mississippi Map Turtle

Life History: There are no data on the habits of this species in Kentucky, and little elsewhere. Apparently egg laying occurs in June. Adults feed principally on vegetation, but damsel- and dragon-fly nymphs, a small clam, snails, and a blue-tailed skink are known to have been eaten by the young.

Painted Turtle XXXVII

Chrysemys picta (Schneider)

Recognition: A small to medium-sized dark turtle having a smooth, unkeeled carapace with an attractive pattern of red or yellow, especially on the marginals. Rear of carapace is not serrate. Head with 2 yellow spots on each side as well as assorted lines and stripes. Carapace length usually 4 to 5½ inches, to 7⅜. Carapace olive to black; plastron plain yellow or with a large oval, dusky, central figure.

Hatchlings are similar in color and pattern to adults but have a strong keel on the carapace, and often the entire plastron is washed with red. They are much more nearly round than adults.

This species is particularly susceptible to the deposition of a reddish incrustation on the carapace and plastron. It can be removed with a finger nail and should not be confused with the normal markings.

Variation: Extremely variable across the state. Two well marked subspecies are recognized:

Chrysemys picta marginata (Agassiz). Midland Painted Turtle. Carapace as described above; large scutes of back alternating instead of continuing in one line across the back. Plastron with a variable dark blotch that is typically oval, involving nearly all the scutes and about half as wide as the plastron or often less. Statewide except in the Jackson Purchase; intergrades along the Cumberland and Tennessee rivers in western Kentucky with:

Chrysemys picta dorsalis (Agassiz). Southern Painted Turtle. Carapace with a broad yellow or sometimes red stripe running down the middle; plastron usually plain yellow, but sometimes with small dusky figures. Occurs in the Jackson Purchase; intergrades with the above.

Males are smaller than females and have extremely elongated front claws when mature. Their tails are longer. The anal opening is under the edge of the carapace in the female, beyond it in the male.

Confusing Species: Most often confused with the stinkpot and the mud turtles, but these have only 11 plastral plates and no red patterns on the carapace.

Habitat: This species seems to prefer quiet, shallow water such as ponds, sloughs, shallow lakes, and bayous. They are most common in areas where aquatic vegetation is profuse.

Kentucky Distribution: Statewide. Apparently least common in the Eastern Coal Field and the Blue Grass. Abundant in most of western Kentucky.

Life History: Although sometimes seen crawling along the bottom under the ice, they spend most of the winter buried in the mud at the bottom. They usually go into hibernation in October or November and emerge in March. However, they have been seen abroad in every month of the year save January.

Mating occurs soon after the spring emergence and may continue irregularly through the summer into September. Mating is preceded by a peculiar courtship. The male chooses a female and by swimming rapidly overtakes and passes her; when well in front he turns suddenly to face her, and as she continues forward, swims backward just in front of her. Extending his forelimbs, palms outward, he rapidly caresses her on the chin and sides of the head with the back of his long claws. After a brief period of this, he withdraws his forelimbs, but continues his backward swimming just in front of her, periodically stroking her again and again.

Egg laying normally occurs in late afternoons in June, but may occur in late May or early July. The female sometimes wanders several hundred yards from the water seeking a suitable place to dig her nest, or she may remain quite near the water's

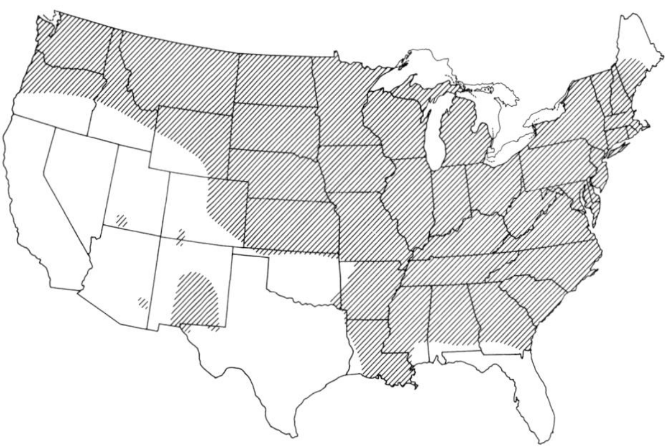

Range of the Painted Turtle in the United States

edge, selecting an open spot where the soil is friable, usually well above the water level. A flask-shaped cavity is dug by scraping first with one hind foot then the other; often the bladder is voided into the partially completed cavity as the body is shifted to bring the other foot into position. The size of the cavity is variable, probably determined by the length of the hind legs. When finished, she deposits from 4 to 10 eggs in the cavity and fills it in with the loose earth excavated from the hole. The neck of the cavity is filled with a kneaded plug of mud, which she flattens and smooths by dragging her plastron back and forth across it. By the time she is finished the nest is well concealed.

The eggs are white, elliptical and, although variable in size, are usually about 1 3/16 x ¾ inches. They sometimes hatch by early September and the youngsters emerge and travel to the water. However, many newly hatched young can be found in early spring, before the eggs for the year are laid. Clearly, these youngsters overwinter in the nest, either as hatchlings or in the egg. Perhaps the warm spring rains are a factor in softening the earth, enabling them to dig their way out.

Little is known of growth and development of the youngsters, but apparently the males reach maturity when about 3½ inches long and 3½ to 4 years old. Females are about 5 inches at maturity, and probably a year older.

Food is variable, but the bulk of the diet is composed of aquatic invertebrates and plants. Neither fishes nor amphibians are regularly eaten, and together probably total less than 2 percent of the total diet.

Remarks: Although essentially aquatic except during the laying season, this species makes rather extensive overland journeys. There is evidence that they sometimes perform migrations in considerable numbers.

Pond Slider

Chrysemys scripta (Schoepff)

Recognition: A large, flattened, green to almost black turtle, usually with numerous longitudinal furrows and whorled figures on the back. There is no keel on the carapace and the rear edge is serrated. Head and neck dark, with stripes, and a light mark behind each eye. Carapace length usually 5 to 8 inches, to 11. Plastron yellowish with dark figures. Under surface of marginals with concentric dark figures.

Young turtles are greenish, with a well-marked contrasting pattern. The shell of the hatchling is almost round and is keeled. With increasing age the carapace becomes blotched with dark spots, often obscuring the original pattern. Sometimes the turtle is almost black, including even the head, neck, and limbs. Males are more susceptible to this creeping melanism than females.

A distinctive characteristic of this species is the rounded lower jaw; it is flat in all other *Chrysemys*.

Variation: Extremely variable across the state; two well-marked subspecies are recognized:

Chrysemys scripta elegans (Wied). Red-eared Turtle. The broad red patch behind each eye is unique among Kentucky turtles; however, the red is sometimes replaced by yellow. The yellow stripes on the legs, neck, and head are narrow; the dark marks under the marginals are wider than the spaces between them. Occurs from the eastern edge of the Knobs south and westward throughout Kentucky except in the Cumberland River drainage of southeastern Kentucky where it intergrades with and is replaced by:

Chrysemys scripta troosti (Holbrook). Cumberland Turtle. Similar to the red-eared turtle but with a narrow yellow spot behind each eye. Yellow lines on legs, neck, and head are broad; the dark concentric spots under the marginals are not so wide as the spaces between them. Occurs in the Cumberland drainage in southeastern Kentucky.

Males are smaller than females and have long, gently curving claws on the front feet. The vent of the male opens beyond the posterior edge of the carapace; in the female it is under the edge.

Confusing Species: Most like the slider and the Missouri slider. The slider has a light C-shaped mark on the second costal scute. The Missouri slider has a plain, or at most lightly marked plastron. In these 2 animals the ventral surface of the lower jaw is flattened.

Habitat: Aquatic, preferring quiet, larger bodies of waters such as coves, sloughs, and bayous; it is often found in farm ponds. Most abundant in shallow water where there is an abundance of aquatic vegetation. Often seen basking on floating or stranded logs, rarely on land.

Kentucky Distribution: From about the eastern edge of the Knobs south and west across the state. Extends eastward in the Cumberland drainage in southeastern Kentucky at least as far as Cumberland, in Harlan County. Most abundant in central and western Kentucky.

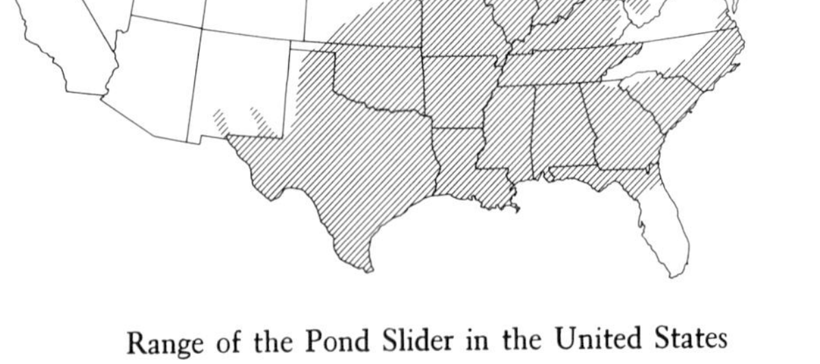

Range of the Pond Slider in the United States

This is one of the commonest turtles sold in the pet trade, and escaped or released animals may be encountered anywhere.

Life History: The winter is spent buried in mud at the bottom of a pond or stream. They usually begin hibernation in October or November, not to emerge until late April or early May, but a warm spell in late winter or early spring may bring them out.

After emergence they begin an elaborate courtship prior to mating, which is remarkably similar to the courtship of the painted turtle. The male swims backward just in front of the female, stroking and gently tapping her about the chin and sides of the face with the backs of the long curved claws.

Egg laying occurs usually in late June and early July. The female leaves the water to search for a suitable spot to dig her nest and lay her eggs. The nest is often located on an embankment, sometimes many yards from the water.

The nest cavity is dug by alternate scraping with the hind feet, and the bladder is frequently voided into the cavity, making the earth more easily dug. The completed cavity is flask-shaped and about as deep as the hind legs can conveniently reach. The bottom of the burrow varies from 3 to 10 inches in diameter. The eggs, numbering from 5 to 22 but averaging about 10, are deposited in the cavity and covered with loose soil. The neck is plugged with a mass of mud and loose material is pulled in and smoothed over the nest, effectively concealing it.

The eggs are white, elliptical, and although variable in size, average nearly 1½ x 1 inches; they have a tough, granular shell.

Although sometimes encountered in the fall, hatchlings are more abundant in the spring. Apparently many spend their first winter in the nest, as baby turtles or as unhatched eggs.

This is another of our turtles whose food habits change with age. Youngsters are largely carnivorous, feeding on a wide variety of invertebrates and smaller vertebrates, but as they grow older, plants constitute a larger proportion of their diet. Older turtles eat crayfish more commonly than any other animal.

Slider

Chrysemys concinna Le Conte

Recognition: A large, flattened, dark turtle with a large, more or less prominent C-shaped mark on the second costal scute. The carapace lacks a keel, is serrated on the rear edge, and is usually "pinched" inward in front of the hind legs. In color it is olive to dark brown, with numerous concentric yellowish lines on the scutes. Carapace length usually 9 to 13 inches, to 14¾. The legs, neck, and head are brown with yellow stripes. The plastron is yellow, usually with a strong dense pattern, especially along the margin of the scutes.

The young are nearly circular in outline, with brighter, more prominent markings and with a keeled carapace.

Variation: No geographic variation is recognized in the state; *Chrysemys concinna hieroglyphica* (Holbrook), the hieroglyphic turtle, is the slider of Kentucky. Males are smaller than females, with much elongated claws on the forefeet and with the vent beyond the back edge of the carapace.

This species hybridizes freely with the Missouri slider, *Chrysemys floridana hoyi*. Identification of these hybrids is difficult, if not impossible.

Confusing Species: Most similar to the Missouri slider and the red-eared and Cumberland turtles. None of these turtles have the C-shaped mark on the second costal, nor do they have a "pinched in" carapace.

Habitat: Aquatic. Usually associated with rivers, it prefers coves, bayous, and sloughs, but is occasionally found in larger ponds and cypress swamps. Often basks on floating or stranded logs.

Kentucky Distribution: Throughout the Jackson Purchase and the Land Between the Lakes. Eastern limits of range unknown, but it has been reported in the lower and middle stretches of

Range of the Slider in the United States

the Green River at least up to Edmonson County. Most abundant in the Jackson Purchase.

Life History: Little data are available on this species in Kentucky. Presumably it lays eggs in June in nests dug near the water. The eggs, numbering up to at least 9 in a clutch, are elliptical, white, and average about 1½ by 1 inch in size.

Juveniles are more carnivorous than adults, feeding on a great variety of invertebrates and smaller vertebrates. Adult diet is largely vegetable matter.

Cooter XXXVIII

Chrysemys floridana Le Conte

Recognition: A large, flattened, dark turtle with a more or less prominent light vertical line (or lines) on the 2nd costal scute. The carapace lacks a keel and the rear edge is serrated. Carapace length 7 to 11 inches usually, up to about 13. Head, neck, and legs striped. Head stripes yellowish, narrow, and broken and twisted. Plastron yellow, usually plain, but may have faint dusky markings. Underside of marginals have conspicuous dark doughnut-shaped (sometimes oval) markings.

Variation: No geographic variation is recognized in Kentucky; *Chrysemys floridana hoyi* (Agassiz), the Missouri slider, is the subspecies of the cooter occurring in the state. Males are smaller than females, with much elongated claws on the forefeet and with the vent beyond the edge of the carapace. This species frequently hybridizes with the slider, *Chrysemys concinna hieroglyphica*, and various intermediates between the two species occur.

Confusing Species: Most like the slider, the Cumberland and the red-eared turtles, but it alone has an unmarked plastron and doughnut marks on the underside of the marginals. Hybrids between the slider and the Missouri slider are difficult to identify with certainty.

Habitat: This is a turtle of the larger, mud-bottomed rivers. It prefers sloughs, oxbows, and backwaters, but is sometimes found in smaller ponds and in swamps.

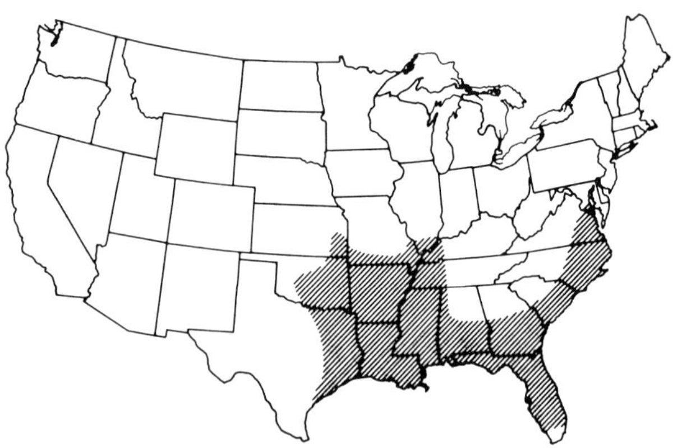

Range of the Cooter

Kentucky Distribution: Occurs throughout the Jackson Purchase in suitable habitats and up the Ohio River probably as far as Daviess County. No data are available on abundance of this species in Kentucky.

Life History: No data are available on the life history of this species in the state, but its habits are likely similar to those of the slider.

FAMILY TRIONYCHIDAE Softshell Turtles

These flattened turtles constitute a small family of aquatic, bottom-dwelling animals found chiefly in rivers and creeks, less often in lakes, ponds, and swamps.

They lack epidermal scales, and their bony shell is much reduced; the plastron is small and lacks a bridge. Softshells are quite flat, and the margins of the spreading carapace are flexible, perhaps an adaptation to bottom dwelling.

The carapace is covered by a leathery skin, and the general appearance of the entire animal is one of softness. The snout is elongated, with the nostrils at the tip. The lips are fleshy but cover a set of sharp, powerful, horny jaws. The limbs are flattened and paddlelike, with only 3 claws on each foot.

Most softshells are carnivorous, lying in wait for their prey or actively foraging. Nearly all are evil-tempered and quick to bite.

Seven genera, arranged in 2 subfamilies are known, occurring in North America, Africa, and Asia. One genus occurs in the United States, and it is represented in Kentucky by 2 species.

1 a. Anterior edge of carapace smooth, without spines; carapace smooth to the touch; legs not patterned; no ridge-like flap inside nostrils / *Trionyx muticus muticus* Le Sueur, Smooth Softshell Turtle (p. 191)

1a

1b

b. Spines present on anterior edge of carapace; carapace granular to the touch; legs heavily patterned; a flap-like ridge on the inside of each nostril / *Trionyx spinifer spinifer* Le Sueur, Eastern Spiny Softshell Turtle (p. 193)

Smooth Softshell Turtle XXXIX

Trionyx muticus Le Sueur

Recognition: A large, aquatic, flattened, leathery-shelled turtle with the anterior edge of the carapace smooth, the legs and feet essentially uniformly colored, and no small ridgelike flap in the nostril. Carapace length: males usually 6 to 6½ inches, to 7; females usually 7 to 13 inches, to 14. Carapace olive-gray, tan, or brown, marked with darker spots or blotches. Plastron much reduced and usually immaculate white. Head, neck, and legs brown, generally unmarked except for a light line from the eye to the throat on each side of the head.

The young have a distinct pattern of dark dashes and a more prominent light margin around the sides and back of the carapace, and the head stripe is more prominent.

Variation: No geographic variation is recognized in Kentucky; *Trionyx muticus muticus* Le Sueur, the common smooth softshell turtle, is the subspecies occurring in the state. Males are smaller than females, and have longer tails. The vent is behind the posterior edge of the carapace in the male, in front of it in the female.

Confusing Species: Most like the eastern spiny softshell turtle, which has prominent tubercles on the anterior edge of the carapace, well-patterned feet, legs, neck, and head, and a ridgelike flap inside each nostril.

Habitat: This species shows a marked preference for clear, sandy-bottomed creeks and rivers; rarely is it found in lakes or ponds.

Kentucky Distribution: Occurs throughout at least the western half of Kentucky in suitable habitats; the eastern limits of the distribution are poorly known. In the north it extends at least to Fleming and Rowan counties, but the eastern limits in the south are unknown.

Life History: This softshell turtle apparently hibernates during the coldest part of the winter, buried in the sand in deep water, emerging in April or early May. Hibernation usually begins in October or November, though they have been seen swimming as late as December.

Mating apparently occurs soon after emergence in the spring, and the eggs are laid in June and July. When ready to lay, a female crawls out on a sandbar with an unobstructed view. Using her hind feet, she scrapes out a hole about 5 inches in diameter and 10 inches deep in the damp sand. After depositing the eggs in the cavity, she covers them with sand and smoothes out the nest site so well that little evidence remains of its location.

The white, spherical eggs have a thick, brittle shell with an average diameter of little over ⅞ of an inch. The clutch usually consists of 18 to 20 eggs but ranges up to 33.

Hatching occurs in about 2 to 2¼ months depending on moisture and temperature. The youngsters emerge at night or early in the morning to make their way to the stream.

This species is almost completely carnivorous. Worms, aquatic insects, snails and other thin-shelled molluscs and crayfish are eaten regularly. Fishes are often taken, and a softshell has been observed (in a large tank) to capture a trout by simply out-maneuvering it. There is evidence that it sometimes preys on young waterfowl.

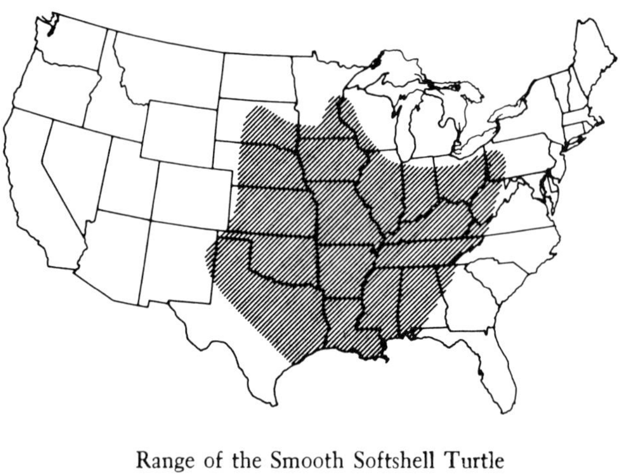

Range of the Smooth Softshell Turtle

Spiny Softshell Turtle

Trionyx spinifer Le Sueur

Recognition: A large, aquatic, flattened, leathery-shelled turtle with the anterior edge of the carapace adorned with tubercles, with the head, neck, legs, and feet conspicuously patterned, and with a small, ridgelike flap just inside each nostril. Carapace length: males usually 5 to 8 inches, to 8½; females usually 6 to 16 inches, to 17. Carapace tan, olive, or dark brown, usually with conspicuous dark spots, blotches, or mottling. Plastron much reduced, immaculate white. Head, neck, legs, and feet olive, with darker stripes and spots. There is a light stripe behind each eye and another on each side of the head and neck starting at the angle of the jaws and extending posteriorly.

The young are paler than adults and have more conspicuous markings. The markings on the carapace are in the form of light-centered, dark-edged spots (ocelli). Males usually retain this juvenile pattern throughout life, but females ordinarily lose the ocelli with increasing age and become marked with irregular dark blotches.

Variation: *Trionyx spinifer spinifer* Le Sueur, the eastern spiny softshell turtle, is the only subspecies known to occur in Kentucky. Although there are insufficient specimens available to draw a definite conclusion, there is evidence that the disposition and size of the markings on the carapace varies across the state, with larger centrally located markings on the softshells from eastern and central Kentucky. In western Kentucky the central and peripheral markings seem more nearly the same size. Probably the differences, if they exist, are insufficient to warrant assigning the turtles to different subspecies.

Males are smaller than females and have longer tails. The vent is located behind the rear edge of the carapace in males, under it in females. Males usually have a rougher, sandpaper-like carapace, especially in the central region.

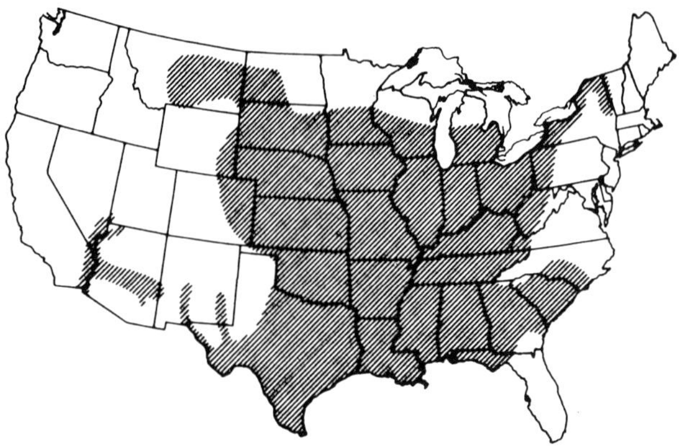

Range of the Spiny Softshell Turtle in the United States

Confusing Species: Most like the smooth softshell turtle, but has tubercles on the front edge of the carapace, patterned soft parts, 2 light stripes on each side of the head, and a ridgelike flap inside each nostril.

Habitat: This species prefers clear, flowing streams and rivers, but occurs in lakes and large ponds. It shows considerably more ecological tolerance than the smooth softshell. Generally, any stream in Kentucky that supports game fish will likely support the spiny softshell. It seems to perfer an area where sand bars or mud flats are available at the water's edge for basking.

Kentucky Distribution: Statewide, but it seems nowhere abundant.

Life History: In late October or November these turtles bury themselves in mud or sand under the water to remain until April or May. Mating occurs shortly after emergence, and egg laying usually follows in late June and continues through July.

When ready to lay, a female searches out an open spot of ground in which to dig a nest. The nest is dug with alternating movements of the hind feet and, when finished, is a flask shaped cavity from 4 to 10 inches deep. The narrow neck is about 1½ inches in diameter and the cavity below is from 3 to 5 inches in diameter. A few eggs are deposited in the bottom of the cavity, covered with soil, and the process repeated until the clutch is complete and the cavity filled. The female then tramples the earth over the nest with the knuckles of the hind feet and reenters the water.

A clutch consists of from about 10 to 25 eggs, with larger females usually laying more eggs. The eggs are spherical, white, thick-shelled but not brittle, and little over an inch in diameter. Often they are slightly out-of-round.

The spiny softshell often basks in the sun on sandbars or mud flats at the water's edge. While basking it is particularly wary and slides quickly into the water and swims away at the least disturbance. One is often seen floating motionless near the surface of the water with the nostrils protruding; such a floating animal can be readily recognized by its pancakelike shape.

Often the softshell will settle on the bottom and with a shuffling motion stir up a cloud of mud, sand, and debris. When this settles, it is almost perfectly concealed from view but can grab any animal that comes within reach. In spite of this predilection for ambushing its prey, the softshell often actively chases down its food.

Basically carnivorous, eating mostly aquatic insects, crayfish, snails, and fishes, the softshell sometimes feeds on vegetable matter. One individual in Ohio had its stomach filled with corn, and watermelon rinds have been successfully used as bait.

ORDER SQUAMATA: SUBORDER SAURIA Lizards

The lizards constitute a diverse group of reptiles, numbering nearly 3,000 species. Their kinship with the snakes is obvious, and these latter have almost surely been derived from lizard-like ancestors.

In the lizards the two halves of the lower jaw are firmly fixed together at the symphysis, and the size of the mouth opening is accordingly restricted. Most lizards have 4 limbs, but some have lost one or both pairs, although traces of their supporting girdles almost always remain. Lizards ordinarily possess external ear openings, movable eyelids, and a third lid, the nictitating membrane. The tail of many lizards is readily detachable and can be regenerated; the vent is a transverse slit.

Saurians are most abundant in the tropics but are found on all continents except Antarctica; in Europe they have extended northward to the Arctic Circle.

Within this range they have adapted to a variety of habitats; deserts, prairies, woodlands, trees, freshwater, the seas, even holes in the earth—all are occupied by lizards. Some have webbed or ridged feet for running over sand or clinging to smooth surfaces. Some of the latter now reside in the dwellings of man and perform some remarkable feats. It is nothing short of amazing to watch the little house geckos of Southeast Asia rush about upside down on the ceiling, catching insects attracted to the lights. Sometimes two or more catch the same insect, and a furious struggle ensues, pulling and tugging, all while the lizards are upside down on a smoothly painted ceiling!

Some of the ground dwellers are large beasts; the heavy-bodied komodo dragon of Indonesia, 10 feet or more long, is the largest. Others are tiny, and some are able to rise up and run rapidly on two legs, much as some of the ancient dinosaurs must have done. Others have completely lost their limbs and slither about like snakes.

The butterflies of the lizard world are the flying lizards, *Draco volans*, of Southeast Asia. These colorful creatures have a folding, winglike flap on either side of the body supported by the greatly elongated last few ribs. When folded against the sides, the "wings" are scarcely noticeable but are colorful and beautifully patterned when spread in "flight." Of course these lizards cannot truly fly, but by spreading the membranes they can glide downward at a fairly flat angle for distances of up to 50 or 60 feet with sufficient maneuverability to catch unsuspecting flying insects as they glide from one tree to another.

There are only two known venomous lizards, the Gila monster and the closely related Mexican beaded lizard. They occur only in the southwestern United States and adjacent Mexico. Many species will bite, however, if given the opportunity.

Twenty-three families of lizards are currently recognized, 9 of which occur in the United States. Four families occur in Kentucky, 3 of them with but a single species each; the other is represented by 5 species in 2 genera.

KEY TO THE FAMILIES OF KENTUCKY SAURIA

1 a. Legs absent / Family ANGUIDAE, Anguids (p. 219)

b. Legs present 2

1a

2 a. Dorsal scales strongly keeled / Family IGUANIDAE, Iguanas (p. 198)

b. Dorsal scales not keeled, smooth or at most granular 3

2a

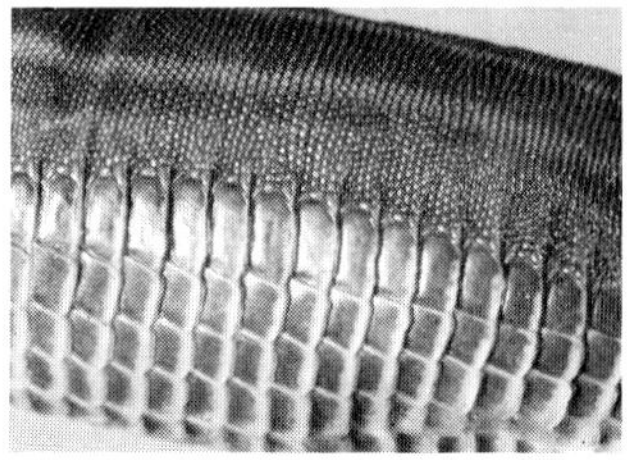

3b

3 a. Dorsal and ventral scales essentially the same size / Family SCINCIDAE, Skinks (p. 201)

b. Ventral scales enlarged, somewhat larger than dorsal scales / Family TEIIDAE, Teiids (p. 215)

FAMILY IGUANIDAE Iguanas

These are the animals most people think of as lizards. About half the genera and three-fifths of the species of lizards in the United States are members of this family. By far the largest family of lizards in the New World, it is almost completely unknown elsewhere. Two genera occur on the island of Madagascar, and one genus inhabits the Polynesian Islands of Fiji and Tonga; otherwise, the family is restricted to the New World.

Iguanas range in size from tiny adults less than 5 inches long to veritable giants of over 6 feet. The smaller iguanids are usually insectivorous or carnivorous, but most of the larger ones are herbivorous. The big marine iguanas of the Galapagos Islands enter the sea to graze on underwater seaweed.

Most iguanas are oviparous and bury their leathery-shelled

eggs in the ground; even the arboreal ones forsake the trees during the laying season. Some few are ovo-viviparous; the females retain the eggs within their bodies until they hatch and then give birth to the hatchlings.

The Iguanidae are represented in Kentucky by a single species.

Fence Lizard XL

Sceloporus undulatus (Latreille)

Recognition: A small to medium-sized gray or brown lizard showing a marked arboreal tendency. The dorsal scales are keeled and pointed, giving the animal a spiny appearance. Total length usually 4 to 7 inches, to 7¼. Dorsal surface sometimes gray and strongly patterned, sometimes brown and less conspicuously patterned. Belly various, from white with dark flecks to conspicuously patterned with blue and black.

Variation: No geographic variation has been observed in Kentucky; *Sceloporus undulatus hyacinthinus* (Green), the northern fence lizard, is the only subspecies occurring in the state. The sexes are quite dissimilar externally. Males: usually brown on the back, with an indistinct pattern; belly, bright blue on either side bordered by black toward the midline; throat with a black-bordered blue patch. Females: usually grayish with a distinct dorsal pattern of wavy dark crossbands, yellowish or reddish at base of the tail; belly and throat whitish, with dark flecks; sometimes with a small amount of pale blue at edges of belly and throat. The young are patterned like the female but darker and duller.

Confusing Species: Our only spiny lizard, all our other lizards are smooth scaled.

Habitat: Fence lizards prefer dry, open, sunny woodlands, usually on hillsides. They may be found in a variety of other sites, however, such as old houses and barns, fence posts, piles

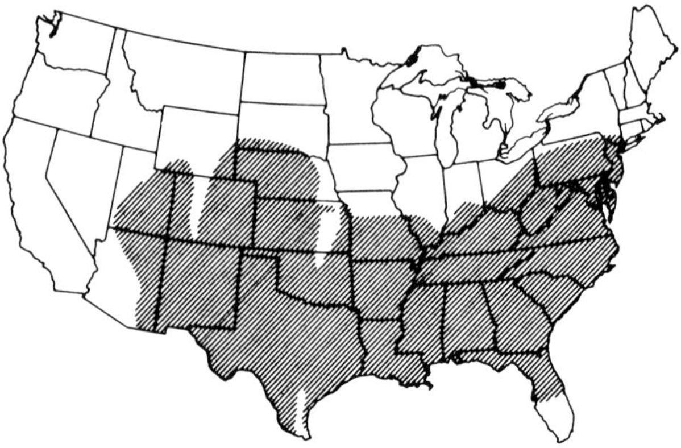

Range of the Fence Lizard in the United States

of logs, rock piles, or similar places. One requirement seems to be a place to climb.

Kentucky Distribution: Statewide. The species is particularly abundant on the dry ridges of eastern Kentucky and on the low hills of western Kentucky.

Life History: Emergence from hibernation occurs from late February to well into March or even April, depending on local weather conditions. Shortly after, breeding activities begin. Males set up small territories and defend them vigorously against all other males. One or two females may usually be found within a given territory, but they do not set up territories of their own. Males apparently make no effort to prevent the females from wandering in and out of any particular territory.

Mating occurs at intervals in April and May, but little is known of the permanence of the attachments formed during this time. The eggs, usually numbering from 6 to 15, are laid from late May to early August, but mostly in a 6-week period around the first of July. When ready to lay, the female selects a site in some open area, in well-drained, preferably sandy soil. There she digs a burrow some 2 to 4 inches deep, depositing her eggs and covering them carefully with the excavated earth. Frequently the middle of a little-used country road is selected as a nest site. After covering the eggs, the female departs and thereafter pays no attention to either the eggs or the eventual youngsters.

After about 10 weeks, usually from late August to mid-September, the eggs hatch and the youngsters dig their way out of the ground. A great deal of growth occurs between hatching and the onset of hibernation in late October or early November. The first year the young grow rapidly, and they are mature by the beginning of their second full season.

Fence lizards will apparently eat any insect or spider that can be captured and swallowed. Sometimes the stomach is literally stuffed with hundred of ants, tiny spiders, and assorted beetles.

Eyesight apparently is quite keen; the fence lizards are able to distinguish slight movements at a considerable distance and capture their prey by a sudden dash and kill it by crushing or by a vigorous shaking. Food is thoroughly chewed, and this habit renders stomach analyses difficult.

FAMILY SCINCIDAE Skinks

Skinks are small secretive lizards, usually with highly polished scales. A few species, however, have spiny scales; one in Australia resembles an animated pine cone. The tail is fragile, readily breaking away but soon regenerated. Many genera have much reduced limbs, and many species are completely limbless or have only vestiges of limbs, a feature that seems characteristic of the family.

Most species bear living young. Usually the eggs are retained in the body of the mother until they hatch, and then the young are born (ovo-viviparity), but a few species exhibit true viviparity, with placenta formation.

The vast majority of skinks are terrestrial, diurnal, and highly active animals, but a few depart from the norm. Some are arboreal but not as much so as some of the iguanas, while some are confirmed burrowers.

Cosmopolitan in distribution, they are more abundant in Africa, the Orient, Australia, and Melanesia, but poorly represented in the United States. The whole Western Hemisphere has only some 5 genera, compared to about 40 in the Eastern Hemisphere.

In Kentucky the family is represented by 5 species, arranged in 2 genera. All have legs, and none bear living young.

KEY TO THE GENERA AND SPECIES OF KENTUCKY SCINCIDAE

1 a. Lower eyelid with a transparent window; dorso-lateral stripes dark / *Lygosoma laterale* (Say), Ground Skink (p. 204)

b. Lower eyelid without a transparent window; dorso-lateral stripes present or absent, but light if present 2

1a(1)

1a(2)

2 a. With a single postmental scale; four light lines on trunk (no mid-dorsal light line); no light lines on top of head / *Eumeces anthracinus anthracinus* (Baird), Northern Coal Skink (p. 213)

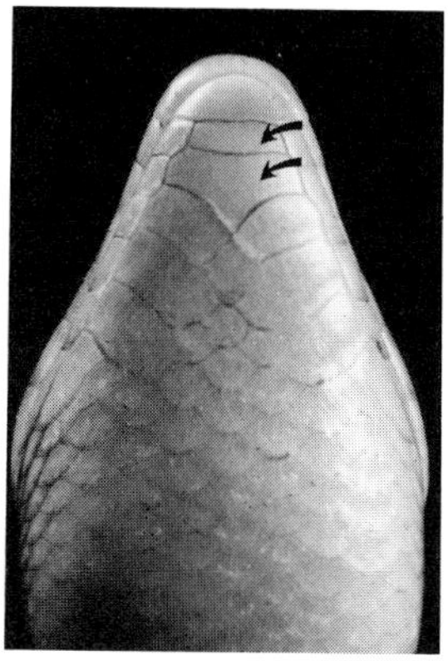

2b

b. With 2 postmental scales; light lines on trunk present or absent, but if present, then 5, and extending onto top of head 3

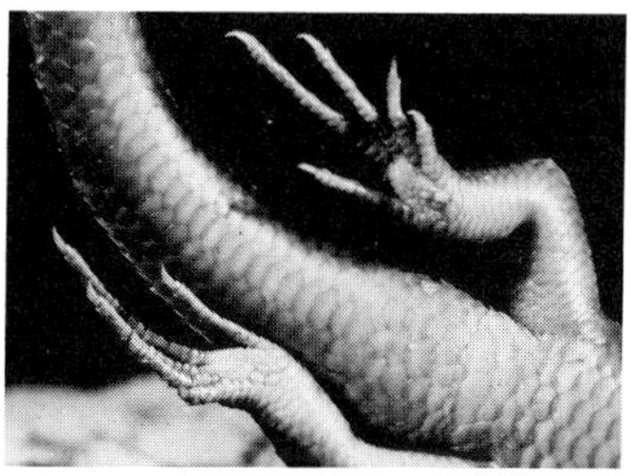

3a

3b

3 a. Scales in middle row under base of tail same size as scales in other rows / *Eumeces inexpectatus* Taylor, Southeastern Five-lined Skink (p. 212)

b. Scales in middle row under base of tail larger than scales in other rows 4

4 a. With 2 enlarged postlabial scales; usually 4 labials anterior to the subocular; 26 to 30 longitudinal rows of

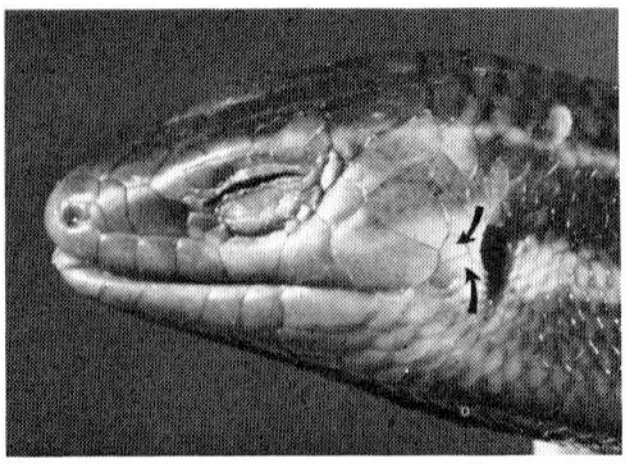

4a

4b

scales at midbody / *Eumeces fasciatus* (Linnaeus), Five-lined Skink (p. 207)

b. With no enlarged postlabial scales; usually 5 labials anterior to the subocular; 30 to 32 longitudinal rows of scales at midbody / *Eumeces laticeps* (Schneider), Broad-headed Skink (p. 210)

Ground Skink XL

Lygosoma laterale (Say)

Recognition: A small, short-limbed, slender, smooth, brown lizard with a dark dorso-lateral stripe. There is a transparent disc in the lower eyelid through which the animal can see when the eye is closed. Length usually 2½ to 4 inches, to nearly 5. The dorsal coloration varies from a light golden-brown or even reddish to a dark chocolate brown. In darker individuals the dark dorso-lateral stripe tends to blend with the dorsal color but is nearly always discernible. The belly is white or yellowish. Newly hatched youngsters have proportionally shorter tails and longer legs than adults, but these differences soon disappear.

Variation: No geographic variation is recognized. Males have heavier tails than females, but there are no reliable external features for separating the sexes.

Confusing Species: Unlike any other lizard; most likely to be confused with one of the plethodontid salamanders, none of which have scales or claws and all of which have a moist skin. Even casual observation will quickly distinguish the skink from the salamanders.

Habitat: A terrestrial species, most often found in wooded areas in the vicinity of streams, seeming to prefer a moist but not wet environment. They are sometimes found in grassy or weedy fields but this seems exceptional.

Kentucky Distribution: Statewide, but rare, if not completely absent from the Blue Grass and the Knobs. They may be found, however, at least along the wooded bluffs of the Kentucky River as it traverses the area. They are abundant in most areas of the state, particularly so in the Eastern Coal Field and the Jackson Purchase.

Life History: The ground skink winters in nooks and crannies below frost line, to emerge with the first warm days of spring. They are not usually out in numbers until April but have been observed as early as February.

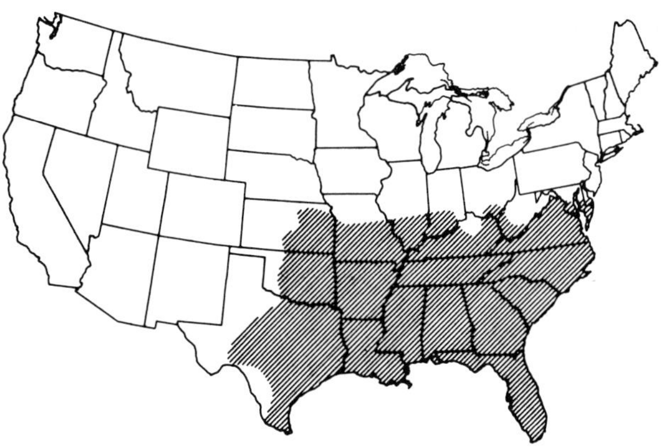

Range of the Ground Skink in the United States

They are shy, secretive animals and so inconspicuous and easily alarmed that, although abundant in most sections of the state, they are seldom seen except by those who know how to find them. When disturbed, they scurry over the leaves of the forest floor for a foot or so with a serpentine twisting gait before burrowing into the leaf litter to hide and lie quiet until danger passes.

Egg laying begins by early June and continues until July or August. By late July, most of the females have deposited their eggs in cavities in rotten logs, stumps, or other humus. An average clutch consists of 3 eggs, but the number varies from 1 to 5. The eggs are elliptical, about 3/16 x ⅜ inches, with thin, white, leathery shells and begin hatching by late July. By late August hatchlings are abundant; some particularly late nests may not hatch until September.

The ground skink is predatory, actively foraging on the forest floor for its food, which consists of small invertebrates. Worms, millipedes, centipedes, spiders, pillbugs, and insects are taken indiscriminately.

Remarks: This is the only lizard in Kentucky with a "window" in the eyelid; it is the only legged lizard in the United States so equipped.

There is a simple technique by which one can see a ground skink almost at will. Walk silently and slowly (at a rate of about 30 or fewer steps per minute) along a woodland path on a sunny spring or summer day. Listen carefully for the faint rustling of the leaves as the skinks scurry over them. By going slowly and silently, one can approach near enough to hear the animals and often see them before they disappear under the leaf mold. It is not uncommon for a particularly alert person to see a dozen or so within a space of a few hundred yards in some areas. The number and variety of other things one can see on such a stroll is remarkable.

Five-lined Skink

Eumeces fasciatus (Linnaeus)

Recognition: This species and the two immediately following (broad-headed and southeastern five-lined skinks) are quite similar and undergo the same changes in color and pattern with increasing age. Juveniles and young adults have bright blue tails and a series of 5 white or yellowish lines running longitudinally on the body; one is in the midline of the back, and two are on each side. All are separated from each other by a darker ground color. With increasing age, the tail gradually loses the blue color, the stripes darken, and the ground color lightens, until old individuals are nearly uniformly colored. Females usually retain some traces of the light lines, but old males sometimes lose them completely, becoming a uniform olive or brown on the trunk and tail. With increasing age the males develop a red pigment on the head during the breeding season, and old males sometimes have completely red heads.

Since these three species are so similar in color, pattern, and the changes associated with sex and increasing age, we must turn to other characteristics to identify them. Fortunately, there are consistent differences in scalation that will separate the three, regardless of the color pattern. In examining the scales on the tail, be sure to consider only the scales on the original, and not a regenerated section.

Eumeces fasciatus has 2 postmental scales, the scales in the middle row under the base of the tail are larger than those in the other rows, and there are 2 enlarged postlabial scales. There are usually 4 labials anterior to the subocular, and 26 to 30 longitudinal rows of scales at midbody. Length usually 5 to 7 inches, to 7½. The dorso-lateral light lines, if present, are on the 4th, or the 3rd and 4th scale rows counting from the midline of the back.

Variation: No geographic variation is recognized. Very variable, color and pattern changing with age, and different in the sexes. See under *Recognition* above.

Confusing Species: Most like the broad-headed skink, which has no enlarged postlabials, 30 to 32 scale rows at midbody, and usually 5 labials anterior to the subocular. Like the southeastern five-lined skink, which has the middle row of scales under base of tail like the other scales. Resembles the coal skink, which has only 4 stripes and only one postmental. Might be confused with the ground skink, which has a dark dorso-lateral stripe and a reddish or brown back.

Habitat: Inhabits a wide variety of situations, but seems most abundant about old houses and barns. Cutover woods with an abundance of rotten stumps and piles of debris usually harbor a good population. Old sawdust piles and overgrown piles of rock seem favored haunts. It usually seeks a slightly damp environment and seems partial to valleys and lower slopes rather than high, dry ridges. It is usually encountered on or near the ground but often climbs about on dead trees and old buildings.

Kentucky Distribution: Statewide; everywhere common in suitable habitats.

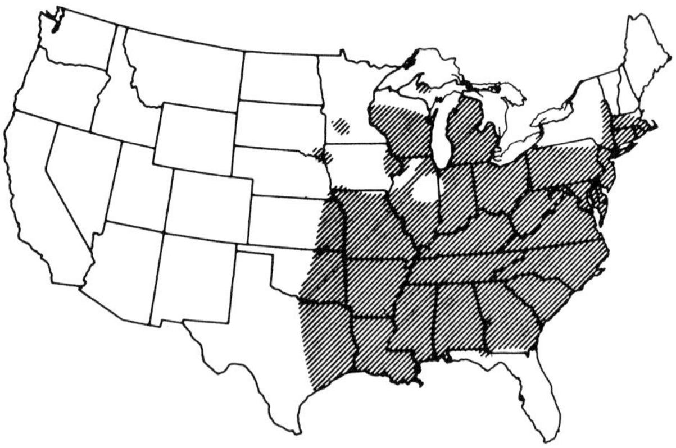

Range of the Five-lined Skink

Life History: Five-lined skinks spend the winter in burrows in the earth, in rock crevices, under large rocks, or similar situations below the frost line. They are among the first lizards to emerge in Kentucky, a series of half-dozen or so consecutive days in March with temperatures in the sixties usually bringing them out to lie in the sun. A subsequent cold spell will drive them back underground, but they generally remain active so long as the temperature does not fall below the low forties. By late April all the skinks are ordinarily out and active.

Courtship and mating usually occur in early May, with the males actively seeking and pursuing the females. Apparently the red head of the male serves as a means by which sex recognition is accomplished at a distance. Males are aggressive, and bloody battles often result when they meet in the breeding season. At other times, when their heads are not red, they seldom fight.

Egg laying follows some 6 to 7 weeks after mating, usually during a 2-week period in June. The female selects an area in or under a rotten log or beneath a large stone and excavates a flask shaped burrow in which she deposits from 4 to 20 but usually 9 to 12 eggs. Often several females nest in the same log or under the same rock. The female remains with the eggs until they hatch, actively defending them from intruders. That her presence with the eggs is not to incubate them is indicated by the fact that the eggs often hatch just the same if the female is removed. She does, however, protect the eggs from damage by intruders, turns them frequently, preventing their adherence to the walls of the cavity, and keeps them moist by voiding the contents of her cloacal bladder in the cavity.

The eggs look like minute chicken eggs. They are about ½ x 5/16 inches when laid but increase their diameter by nearly 40 percent by hatching time. They have a tough, flexible, although thin leathery shell. When dropped, they often bounce.

Hatching occurs usually from early July to mid-August after an incubation period of from 27 to 47 days, varying with temperature and humidity. After hatching, the youngsters usually remain in the nest cavity for a day or so with the mother curled protectively about them. When they disperse, all family ties are broken permanently.

In late summer skinks become increasingly secretive and are rarely found in October, even when the temperature is ideal and food abounds. Captives become listless about this time, and it seems that the wild ones must simply be in hiding.

An efficient predator, the five-lined skink feeds largely on assorted invertebrate animals. Spiders are a mainstay, and many kinds of insects are taken; small snails are an important part of the diet. Sometimes an adult will feed on other lizards or baby mice; rarely one feeds on the eggs or young of its own species.

Broad-headed Skink XLI

Eumeces laticeps (Schneider)

Recognition: For a discussion of the 3 species of "five-lined" skinks in Kentucky, see *Recognition* under the five-lined skink, *Eumeces fasciatus*.

The broad-headed skink has 2 postmental scales, and the scales in the middle row under the base of the tail are larger than those in the other rows. There are no enlarged postlabial scales, and there are usually 5 labials anterior to the subocular. The longitudinal rows of scales at midbody number 30 to 32. Length usually 6 to 12 inches, to 12⅜. Dorso-lateral light lines, if present, on the 4th, or 3rd and 4th scale rows counting from the midline of the back. This is our most massive lizard; any red-headed lizard over 9 inches in total length is almost surely this species.

Variation: No geographic variation is recognized. Color and pattern vary with age, sex, and season as with the five-lined skink.

Confusing Species: Most like the five-lined skink, which is often smaller, has 2 enlarged postlabials, usually 4 labials anterior to the subocular, and 26 to 30 scale rows at midbody. Similar to the coal skink, which has 4 stripes and only one postmental scale.

Habitat: Woodlands. The most arboreal of our skinks, it often climbs high into trees and hides in hollow trees and knot holes. Although sometimes found in low, swampy woods, it prefers more upland areas, where it is more often found on the drier ridges and hillsides than in the valleys.

Kentucky Distribution: Probably occurs throughout Kentucky, but records are surprisingly few for so conspicuous an animal. It seems everywhere rare, except perhaps in the Mammoth Cave area.

Life History: There are little data on the life history of this species in Kentucky. Eggs are laid in late June or early July, in clutches of 6 to 10, in cavities in stumps or hollow trees. The female presumably stays with the eggs until they hatch.

Remarks: Any accurate observations on this species in Kentucky would be worthwhile.

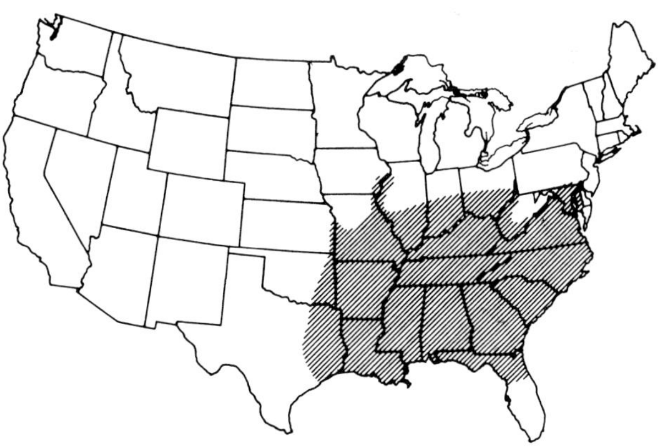

Range of the Broad-headed Skink

Southeastern Five-lined Skink

Eumeces inexpectatus Taylor

Recognition: For a description of the 3 species of "five-lined" skinks in Kentucky, see *Recognition* under the five-lined skink, *Eumeces fasciatus*.

The southeastern five-lined skink has 2 postmental scales, and the scales in the middle row under the base of the tail are essentially the same size as those in the other rows. There are 2 enlarged postlabial scales. Length usually 5 to 8 inches, to 8½. Dorso-lateral light lines, if present, on the 5th, or the 4th and 5th scale rows counting from the midline of the back.

Variation: No geographic variation is recognized. Variations correlated with age, sex, and season as in the five-lined skink.

Confusing Species: Most like the five-lined skink, which has the scales in the middle row under the base of the tail larger than in the other rows. Like the broad-headed skink, which also has the middle scales under the tail enlarged, has no enlarged postlabials, usually 5 labials anterior to the subocular, and 30 to 32 longitudinal scale rows at midbody. Similar to

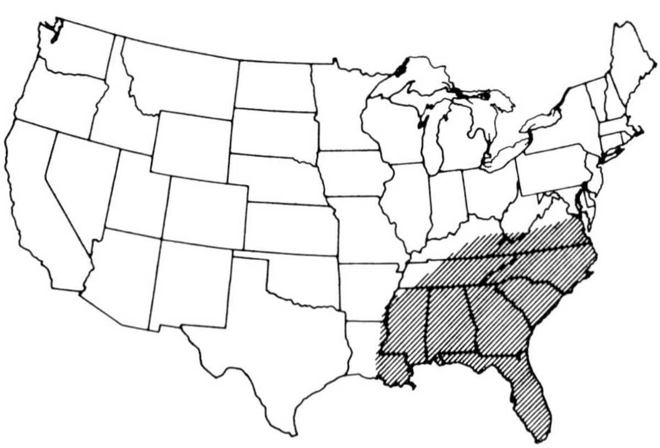

Range of the Southeastern Five-lined Skink

the coal skink, which has only 4 stripes and one postmental scale.

Habitat: Usually found in or near woodlands, especially cut-over woods, with numerous stumps, piles of brush, and other debris. Old buildings, fences, and unused sawmill sites are often utilized. Commonly encountered on the ground under logs, boards, or stones, but it climbs readily. This species is able to survive in drier areas than either of our other "five-lined" skinks.

Kentucky Distribution: Poorly known. Occurs in southeastern Kentucky, as far west as Barren County and as far north as Powell County. It should be looked for in southeastern Kentucky, especially south of a line connecting Boyd and Todd counties.

Life History: No data are available for this species in Kentucky; likely its life history is similar to that of the five-lined skink.

Remarks: Accurate observations of this species would be valuable.

Coal Skink XLII

Eumeces anthracinus (Baird)

Recognition: A medium-sized, smooth, glossy lizard with a broad median dorsal stripe and 2 light lines on each side of body. There are no light lines on the top of the head, and there is but a single postmental scale. Length usually 4½ to 6½ inches, to 7. Color olive gray to brown on the back, in the form of a band from the head to near the tail tip. On either side of this band is a narrow dorso-lateral white line. There is a broad, rich brown to black stripe along the side below the

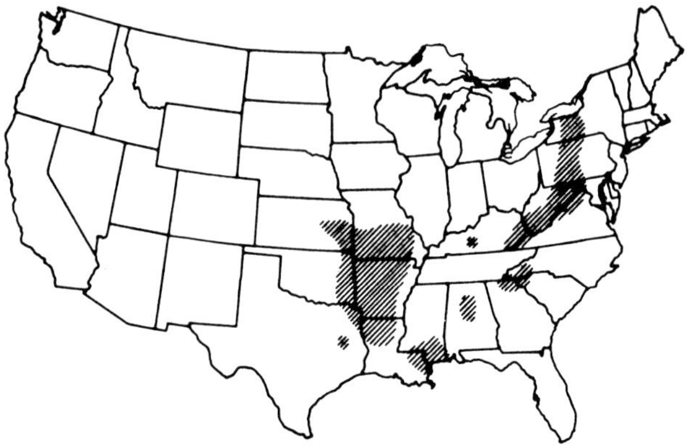

Range of the Coal Skink

dorso-lateral line, bordered below by a narrow white line on the side of the belly, from ear to groin. Below this second narrow light line is a grayish area that becomes lighter ventrally until it fades imperceptibly into the gray or bluish belly.

Normally there is no mid-dorsal light line, but in some individuals there is an incomplete, narrow light line on the neck; if present, such a mid-dorsal light line never extends onto the head.

Variation: There are so few Kentucky specimens of the coal skink that no geographic variation has been recognized; *Eumeces anthracinus anthracinus* (Baird), the northern coal skink, is the only subspecies recorded from the state. It is possible that coal skinks of western Kentucky are sufficiently like a more westerly race as to be identified with them rather than the northern form. Resolution of the problem must wait additional data.

Youngsters are often marked like the adults but are sometimes uniformly dark, even nearly black; however, they have a deep blue-violet tail.

Confusing Species: Most often confused with the 3 "five-lined" skinks. In these 3, if the light lines are visible, the mid-dorsal one is well developed and continues forward onto the head; if the stripes have disappeared, check for 2 postmentals.

Habitat: A woodland species, usually found in a chestnut oak-poplar association. They secrete themselves in rotten logs, stumps, brush piles, or beneath stones, and sometimes may be found beneath scrap boards and slabs about a deserted sawmill site. Apparently they prefer humid habitats that are not far removed from water.

Kentucky Distribution: Poorly known; there are records from Bell, Harlan, Breathitt, and Edmonson counties. One of the rarest reptiles in the state.

Life History: There are no data for this species in Kentucky. In New York, courtship and mating occur in late May, and 8 or 9 eggs are laid in June and guarded by the female. The eggs are about ¼ x ⅜ inches when laid, but increase in size with time. They hatch in 4 to 5 weeks, and the newly hatched youngsters are about 2 inches long.

Remarks: Accurate observations on this rare animal in Kentucky would be valuable.

FAMILY TEIIDAE Teiid Lizards

The Teiidae comprise a New World family of long-tailed, long-tongued lizards. Most have well-developed limbs, but a few are burrowers with degenerate limbs.

Most teiids are terrestrial or fossorial, but at least one is semiaquatic. A few feed on plants, but the great majority are carnivores, actively seeking out and capturing their prey. One species has a highly specialized diet; it feeds on snails.

The Teiidae are restricted to the Western Hemisphere.

About 70 genera are known, and nearly all of them are confined to South America. A single genus, *Cnemidophorus*, reaches the United States; it is represented in Kentucky by a single species.

Six-lined Racerunner XLIII

Cnemidophorus sexlineatus (Linnaeus)

Recognition: An elongate, dull (not shiny), olive, gray, brown, or gray-green lizard with a long tail and 6 longitudinal light stripes on the body. Length usually 5 to 9 inches, to 9½. The 6 light stripes are pale blue to yellowish and extend from on the head well onto the base of the tail. The belly is white or washed with blue. The young have relatively shorter tails and the light stripes are yellow and quite distinct.

The ventral scales are unique among our lizards; there are 8 rows of large, rectangular plates on the belly. The dorsal scales are tiny, and number literally in the thousands. The body, and especially the tail, is rough to the touch.

Variation: No geographic variation is recognized in Kentucky; *Cnemidophorus sexlineatus sexlineatus* (Linnaeus), the common six-lined racerunner, is the only subspecies occurring in the state. Males may be recognized by several external features: the belly is usually washed with blue in the breeding season; the base of the tail is more massive; the femoral pores are considerably more distinct; and the lateral light stripes are less distinct.

Confusing Species: Could be confused only with the skinks, but none of the skinks have 8 rows of enlarged rectangular scales on the belly or as many as 6 light lines on the body.

Habitat: A terrestrial species, preferring areas of sandy or loose soil, where the ground cover is sparse. They avoid damp areas and those covered with heavy grass or other moisture-retaining vegetation.

Kentucky Distribution: Western Kentucky, eastward at least to Warren, Edmonson and Meade counties. The eastern limits of the range are unknown, but it likely does not extend much farther east than Edmonson or Meade County. There is a possibility that a northward extension of the range in east Tennessee extends into Kentucky in the area of Bell County, and it should be looked for at that region. Racerunners are particularly abundant in southern Calloway County.

Life History: These sun-loving lizards do not emerge from hibernation until April or May, depending upon the temperature. Several consecutive sunny warm days in mid-April will bring them out, but they retreat underground with the advent of a cold snap.

Daily activity begins in the morning of warm sunny days and soon reaches a peak; by mid-afternoon they are less active and by late afternoon all have disappeared. They are extremely wary, but with care an observer can approach within 15 or 20 feet without disturbing one.

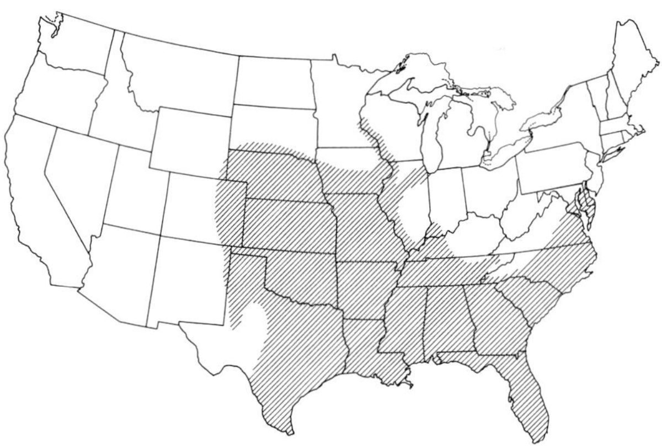

Range of the Six-lined Racerunner in the United States

When disturbed, they dart away to seek shelter and are very difficult to catch. An active lizard can easily outmaneuver the most agile person and handily win a race with most; one has been clocked at 18 miles an hour.

Mating occurs in spring soon after emergence. In an elaborate courtship, the males moves over the ground in a figure-8 path while swinging his hips from side to side. He apparently courts males and females indiscriminately, but when he approaches a male he is attacked.

Between early June and middle July the female lays from 3 to 6 eggs in a cavity some 4 to 12 inches deep, usually hidden under some sheltering object such as a log. The egg is elliptical, about 7/16 x 11/16 inches, with a soft, semitransparent, almost salmon-colored shell when first laid. It is quite smooth, and dirt adheres to it less than to some other species; usually eggs that have been in the nest for some time are still clean and white. Some youngsters are abroad by late July, but most do not appear until August.

Food consists of a variety of invertebrates—ants, flies, spiders, small moths, and various larvae—with grasshoppers and crickets making up a substantial percentage of the diet. Soft-bodied, small insects seem preferred. Beetles are not frequently eaten, and some, such as the lady bird beetles, seem positively distasteful. In any event, after taking one, the lizard promptly rejects it and wipes his mouth on the ground. Larger moths and butterflies, although captured, may not be eaten.

With the cooler days of autumn, racerunners become less active and spend more time underground. They usually disappear for the winter in late October.

FAMILY ANGUIDAE Anguids

This family consists of about 10 genera, largely confined to tropical America; only 2 are present in the Old World. *Ophisaurus* occurs in northern Africa, southern Europe, and Asia, as well as in North America. One genus, *Anguis*, is found only in Europe, North Africa, and Asia.

Most anguids are terrestrial; a few are somewhat aboreal. Most are diurnal, but a few species remain hidden by day to emerge and feed by night. Some have well-developed limbs, some are completely legless.

Most are oviparous, but ovo-viviparous species are known; one genus, *Gerrhonotus*, in the western United States exhibits both means of reproduction.

Anguids are of two basic types. The 2 North American genera possess a prominent groove along each side that is filled in with minute granular scales quite unlike the larger, heavy, platelike fixed scales above and below the groove. The groove extends from the head to near the base of the tail and enables the otherwise heavily armored lizard to expand the body diameter as when a hearty meal is eaten or eggs develop. The remaining genera lack the grooves but are similarly armored.

A single species of the family occurs in Kentucky.

Slender Glass Lizard XLIII

Ophisaurus attenuatus Baird

Recognition: An elongate, shiny, striped, legless lizard. Unlike a snake it has movable eyelids and external ear openings. Length usually 20 to 36 inches, to about 42. Dorsum rich brown, with a dark median stripe or series of dashes and well-marked dorso-lateral dark stripes. There is a series of crossbands down the back. A longitudinal series of alternating light and dark stripes appears on each side, and there are some dark stripes below the longitudinal groove. The young differ from adults in having less well-marked dorsal crossbands and more conspicuous longitudinal stripes.

Variation: No geographic variation is recognized in Kentucky; *Ophisaurus attenuatus longicaudus* McConkey, the eastern slender glass lizard, is the only subspecies occurring in the state. There seem to be no external features by which one can reliably separate the sexes.

Confusing Species: This lizard is most often confused with various snakes; however, no snake has movable eyelids or external ear openings. When held, the glass lizard seems stiff and inflexible, while snakes are flexible and supple.

Habitat: A completely terrestrial lizard, inhabiting grassy fields, brushy areas, and open woodlands and seeming to prefer dry, upland areas. In Edmonson County it is most often encountered in grasslands; farmers sometimes find them while working in grassy hayfields or plowing in such areas. Specimens have been found dead on roads in McCreary County near Cumberland Falls; surely in that area they live in open woodlands.

Kentucky Distribution: Known only from McCreary, Barren, Edmonson, and Hart counties. The species probably occurs as isolated populations throughout the Mississippian Plateau, the Western Coal Field, and the Jackson Purchase. This animal seems rare throughout the state, but because of its secretive habits the rarity may be more apparent than real.

Life History: The glass lizard is a confirmed burrower and apparently spends much of its life underground. There is some question whether the animal is diurnal or nocturnal; the answer awaits more information. Although they have been observed

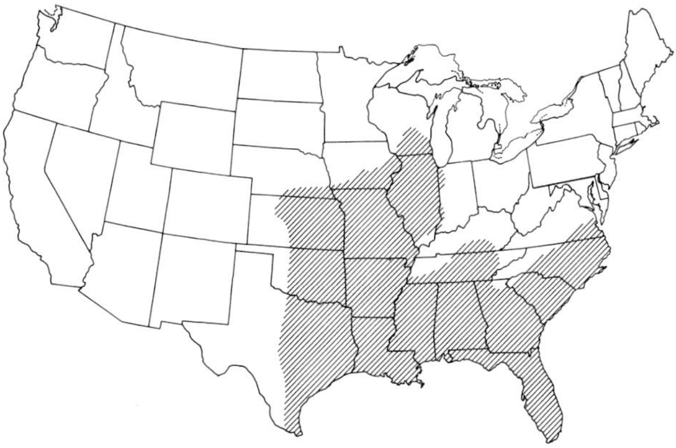

Range of the Slender Glass Lizard in the United States

sunning by day, most individuals encountered by day are either found in hiding or started from such places.

The glass lizard exhibits an adaptation to forestall predation that has given rise to an amazing myth. The tail of the animal is long, sometimes as much as twice the length of the head and body, and when the lizard is grasped or struck the tail often breaks off, sometimes in two or more pieces. These pieces wriggle about, occupying the attention of the predator while the rest of the lizard escapes. This characteristic is responsible for its being erroneously called a "glass snake." A commonly held belief is that the animal can return and rejoin its lost pieces to the body, to crawl off as good as new. Obviously, this is quite impossible. However, the lizard does regenerate a new tail, but it is never as good as the original one. Seldom is a complete glass lizard found; most have lost their tails at some time or another.

Essentially nothing is known of the reproduction of this species in Kentucky. Apparently from 8 to 17 eggs are laid from June to early August in cavities under some sort of shelter, and the female remains with them until they hatch.

The eggs may be pointed at one end and are about ½ x ¾ inches. By hatching time, the eggs have increased to about

¾ x 1 inch. At hatching, the youngsters are from 4⅜ to 5¼ inches long.

Food of the glass lizard consists primarily of arthropods. Many kinds of insects are eaten, as are other invertebrates. They sometimes take small vertebrates, such as lizards and snakes, and in captivity are sometimes cannibalistic.

ORDER SQUAMATA: SUBORDER SERPENTES
Snakes

There are approximately 2,700 species of snakes distributed throughout the world. They have been the objects of more interest, study, and speculation than any other single group of reptiles or amphibians and probably more than all other groups combined. Although they have undergone considerable adaptation to different habitats, they all possess certain characters in common.

Snakes are elongate animals with no apparent limbs or limb girdles. Some few retain vestigial pelvic girdles and tiny spurlike remnants of hind limbs. The sternum is absent; hence all the ribs are floating. The ear is much reduced, and snakes cannot hear in the sense that man does. There is no external ear opening, tympanic membrane (eardrum), middle ear, or Eustachian tube.

Except in a few burrowing species the eye is covered with a fixed transparent scale beneath which the eyeball moves; our snakes cannot close their eyes. When the skin is shed, the transparent scale covering the eye is shed just like the rest of the skin, and the new one beneath is exposed.

The tongue is long, protractile, and forked; within limits the individual forks can be controlled independently. It is used solely as an organ of taste and smell and, contrary to popular belief, is not a stinger of any sort. The touch of a snake's tongue is so gentle as to be barely felt.

The construction of the jaws is quite distinctive. The halves of the lower jaw are attached to each other at the chin by an elastic ligament and attached to the skull posteriorly by a loose double joint that allows them to be moved independently in nearly any direction—forward, back, in, or out. The upper jaws are also loosely attached to the skull and capable of independent movement. In addition, the body is quite distensible; consequently snakes can swallow objects substantially larger than their usual body diameter. When swallowing a large object, a snake can be seen to "walk" its mouth about the object by moving first one half of each jaw forward, then the other; it seems to literally crawl about its prey.

As a group, the Serpentes are widely distributed, occupying suitable habitats on all continents save Antarctica and such islands as New Zealand and Ireland. They are most abundant in tropical countries, especially in the Old World, and become less abundant in the temperate regions both north and south.

They occupy a variety of habitats. Most are terrestrial; some are fossorial, burrowing beneath the surface; some arboreal, rarely if ever coming down from the trees. Many are aquatic in varying degrees. Some prefer to live in moist places, some spend part of the time in water, but some of the sea snakes are strictly marine, never coming ashore.

Snakes are quite variable in size, ranging from tiny creatures, rarely attaining a length of over 4 inches, to veritable giants, over 30 feet in length and weighing several hundred pounds. The record for length is held by a South American anaconda that was measured on the upper Orinoco River by a petroleum geologist with a steel tape; it was 37½ feet long.

Some snakes are highly venomous, and the death rate from their bites is high in certain parts of the world. In the United States, however, the death rate from snake bite is calculated as between .015 and .02 per 100,000 population. This is substantially less than the death rate from slipping in the bathtub or from being struck by lightning. The toxicity of the venom varies from one snake to another, even within a species. Of the venom of two female timber rattlesnakes, collected in the same place on the same day, that of one was approximately 5 times as toxic as the other. In the United States, rattlesnakes as a group account for the greatest number of bites and fatalities. The copperhead accounts for the most bites, but the bite is almost never fatal, even in an untreated child. The western diamond-backed rattlesnake is second in number of bites but responsible for by far the greatest number of deaths.

The Serpentes are organized into 10 families, all of which occur in the Orient; 3 are restricted to that region. Seven families occur in the New World, but none is restricted there. Five families occur in the United States, but only 2 extend into Kentucky.

KEY TO THE FAMILIES OF KENTUCKY SERPENTES

1 a. Without a pit between the eye and the nostril; pupil of eye round; scutes on the underside of the tail in a double row; teeth all essentially similar, none modified into large fangs / Family Colubridae, Harmless Snakes (p. 225)

b. With a facial pit between the eye and the nostril; pupil elliptical or a vertical slit; scutes on the underside of the tail mostly in a single row; with a pair of large hinged fangs in the upper jaw / Family Viperidae, Subfamily Crotalinae, Pit Vipers (p. 301)

1a

1b

FAMILY COLUBRIDAE Harmless Snakes

By far the great majority of the snakes of the world are members of this large, cosmopolitan family. Most of them are small to moderate in size, and none attain the massive size or have the vestigial hind limbs or pelvic girdles exhibited by the family Boidae, the boas, pythons, and anacondas.

Usually the teeth are solid, and neither grooved nor connected with poison glands. Some few, however, have grooved fangs in the rear of the mouth and supralabial glands modified for venom secretion. Such rear-fanged snakes are unable to inject venom by striking but must inject it by chewing. Obviously such an arrangement is of value only in subduing some already seized prey. One of our smallest colubrids, *Tantilla*, is a member of this group of rear-fanged snakes. It is so small, however, as to be completely harmless to man. A few of the larger rear-fanged snakes in other parts of the world are mildly dangerous; there are some records of fatalities from their bite.

Colubrids occupy essentially all available habitats. Most are terrestrial, many aquatic, some fossorial, and some arboreal. Food is equally diverse. Some feed almost exclusively on birds and mammals; others limit their diet chiefly to amphibians and reptiles. Many are insectivorous; some feed principally on worms. Some are snail eaters; others feed on fish. Many feed on nearly any animal they can manage. In spite of their diversity, they exhibit at least 2 common characteristics regarding their food and feeding: All are predators and all swallow their prey whole.

Oviparous, ovo-viviparous, and viviparous species are all included in the Colubridae, and all three occur in Kentucky. As a general rule, the more aquatic snakes are more likely to produce living young than the more upland forms; however, this is no more than a general rule; one of our most aquatic snakes, *Farancia*, lays eggs; one of our upland snakes, *Storeria*, gives birth to living young.

The family Colubridae is represented in the United States by some 93 species arranged in about 39 genera. Of these, 28 species and 17 genera occur in Kentucky; 10 species are represented by more than one subspecies.

KEY TO THE GENERA AND SPECIES OF KENTUCKY COLUBRIDAE

1 a. A single internasal shield 2
b. Two internasal shields 4

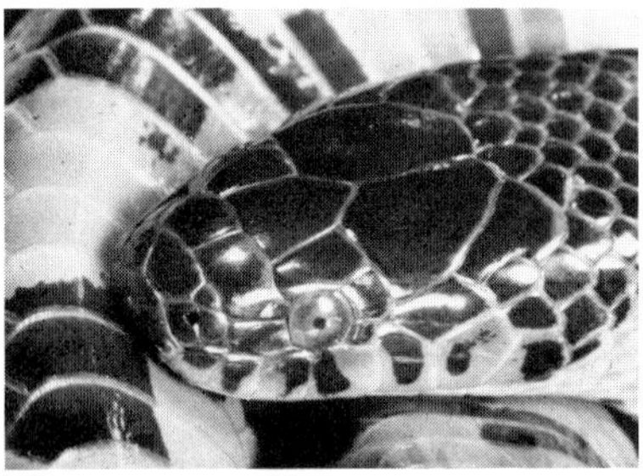

1a

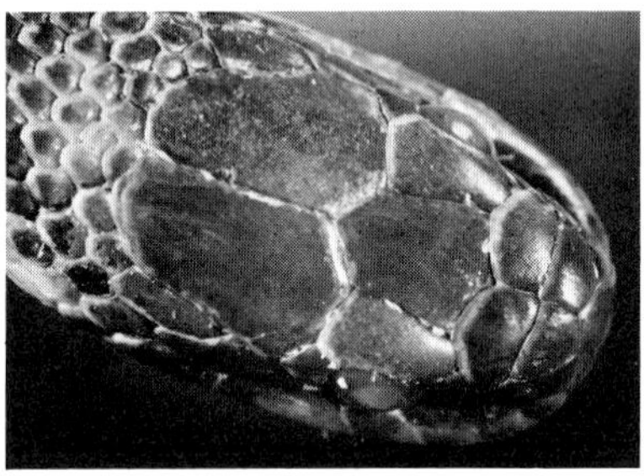

1b

2 a. Dorsum brownish or grayish; belly plain white or yellowish; scales smooth or weakly keeled; upper labials 6, lower labials 6 (*Virginia valeriae*, Earth Snakes) 3
b. Dorsum black; belly red and black; scales smooth;

upper labials 7–8, lower labials 8–10 / *Farancia abacura reinwardti* Schlegel, Western Mud Snake (p. 241)

3 a. Scale rows 15; scales mostly smooth, but usually with faint keels on back near tail; color gray to grayish-brown / *Virginia valeriae valeriae* Baird and Girard, Eastern Earth Snake (p. 281)

b. Scale rows 17; scales weakly keeled; color reddish to grayish-brown / *Virginia valeriae elegans* Kennicott, Western Earth Snake (p. 281)

4a

4b

4 a. Dorsal scales keeled, sometimes only faintly so on uppermost scale rows and on posterior part of body 5

b. Dorsal scales smooth 29

5 a. Anal plate divided 6

b. Anal plate entire, not divided 25

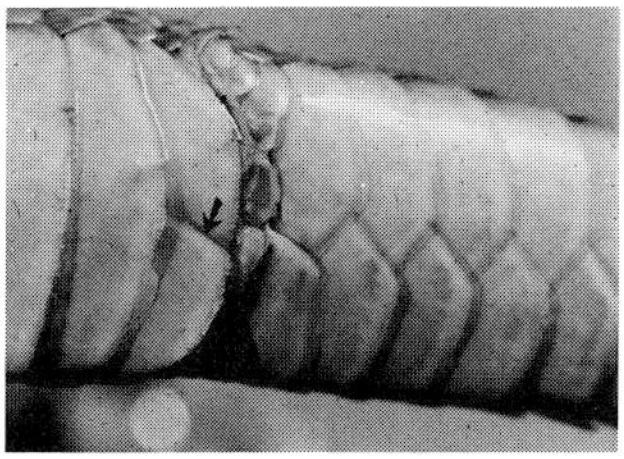

5a

5b

6 a. Rostral turned up and keeled above / *Heterodon platyrhinos platyrhinos* Latreille, Eastern Hognose Snake (p. 246)

b. Rostral normal 7

6a

7 a. No loreal; preocular in contact with posterior part of nasal (Genus *Storeria*) 8

b. Loreal present 10

7a

7b

8 a. Scales in 15 rows; belly usually red; occiput usually with 3 light spots / *Storeria occipitomaculata occipitomaculata* (Storer), Northern Red-bellied Snake (p. 285)

b. Scales in 17 rows; belly not red; occiput unspotted 9

9 a. Dorsal spots distinct, not fusing to form crossbands; ventrals and caudals less than 176 / *Storeria dekayi dekayi* (Holbrook), Northern Brown Snake (p. 283)

b. Dorsal spots fused to form narrow crossbands; ventrals and caudals 176 or more / *Storeria dekayi wrightorum* Trapido, Midland Brown Snake (p. 284)

10 a. No preocular; loreal in contact with the eye (*Virginia valeriae*, Earth Snakes) 3

b. Preoculars present, 1 or 2 separating loreal from the eye 11

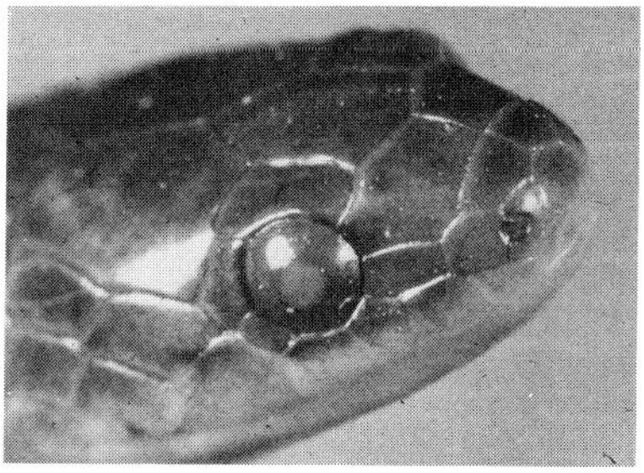

10a

10b

11 a. Scale rows 17; color grass green; dorsum unpatterned / *Opheodrys aestivus aestivus* (Linnaeus), Eastern Rough Green Snake (p. 248)

b. Scale rows more than 17; color not grass green; dorsum patterned or not 12

12 a. Dorsal scales weakly keeled; postoculars 2; belly flat, meeting body at an angle; body shaped like a loaf of bread in cross section (Genus *Elaphe*) 13

b. Dorsal scales strongly keeled; postoculars 2–3; body rounded in cross section 15

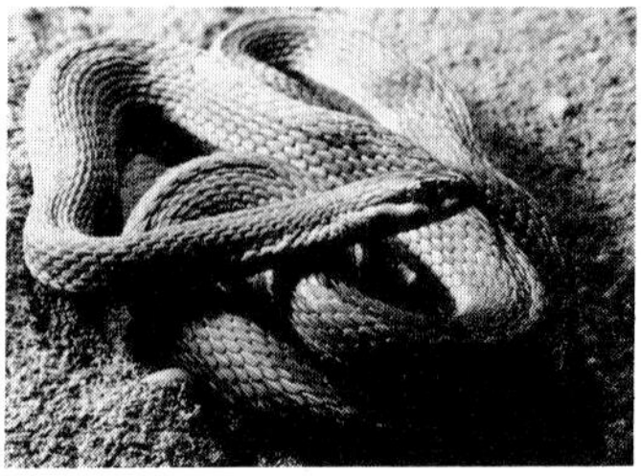

12b

13a

13 a. With 4 anterior stripes, the upper 2 of which traverse the parietals and unite between the eyes / *Elaphe guttata guttata* (Linnaeus), Corn Snake (p. 254)

b. With no such neck stripes (*Elaphe obsoleta*, Rat Snakes) 14

14 a. Dorsal scale rows 25; adults essentially black, sometimes with faint traces of a pattern / *Elaphe obsoleta obsoleta* (Say), Black Rat Snake (p. 256)

b. Dorsal scale rows 25–27; adults with a black pattern on a gray ground color / *Elaphe obsoleta spiloides* (Duméril, Bibron, and Duméril), Gray Rat Snake (p. 256)

15 a. Scale rows 19 at midbody 16

b. Scale rows more than 19 at midbody 17

16 a. Belly yellowish to greenish, often with 2 rows of dark spots down the midline and 3 dark lines down back / *Regina septemvittata septemvittata* (Say), Queen Water Snake (p. 287)

b. Belly red, with row of conspicuous black spots down each side; back with 4 rows of dark blotches arranged in 2 alternating rows on either side of the midline / *Natrix kirtlandi* (Kennicott), Kirtland's Water Snake (p. 293)

17 a. Scale rows 27–33 (except sometimes 25 in *Natrix rhombifera,* but the pattern of alternating dorsal and lateral spots with oblique connecting bars is distinctive); lower labials usually 11–13 18

b. Scale rows 21–25 at midbody; lower labials usually 10 19

18 a. Eye in contact with upper labials / *Natrix rhombifera rhombifera* (Hallowell), Diamond-backed Water Snake (p. 295)

b. Eye separated from upper labials by a series of suboculars / *Natrix cyclopion cyclopion* (Duméril, Bibron, and Duméril), Green Water Snake (p. 289)

18b

19 a. Body essentially uniformly colored above (*Natrix erythrogaster,* Red-bellied Water Snakes) 20

b. Body with crossbands or spots above 21

20 a. Dorsum gray or greenish gray; belly yellow, sometimes washed with orange; ventral scutes occasionally with dark pigment at their ends / *Natrix erythrogaster flavigaster* Conant, Yellow-bellied Water Snake (p. 291)

b. Dorsum dark, sometimes black; belly red; ventral scutes often with dark pigment at their ends / *Natrix erythrogaster neglecta* Conant, Northern Copperbelly (p. 291)

21 a. Belly immaculate or with dark antero-lateral margins of ventrals; tip of tail immaculate beneath 22

b. Belly with well-defined markings; tip of tail not immaculate beneath 23

22 a. Dorsal pattern of dark discrete blotches alternating with a lateral series of dark vertical bars / Juvenile *Natrix erythrogaster flavigaster* Conant, Yellow-bellied Water Snake (p. 291)

b. Dorsal pattern as above but tending to run together; dorsal blotches not discrete / Juvenile *Natrix erythrogaster neglecta* Conant, Northern Copperbelly (p. 291)

23 a. Dorsal pattern wholly of transverse bands or rarely with a few alternating spots; dorsal markings on body usually 11–17; belly with squarish spots / *Natrix fasciata confluens* Blanchard, Broad-banded Water Snake (p. 299)

b. Dorsal pattern of 3–15 anterior transverse bands, posterior to which is a series of median dorsal blotches alternating with a series of lateral spots (rarely a complete series of transverse bands, but then the belly has half-moon-shaped marks (*Natrix sipedon*, Common Water Snakes) 24

24 a. Usually 3–10 anterior transverse bands and a total of more than 30 bands and dorsal spots on body; ventrals 135–155, many with two dark or red half-moon-shaped marks with flecking or mottling; lateral interspaces narrower than lateral bars / *Natrix sipedon sipedon* (Linnaeus), Northern Water Snake (p. 297)

b. Usually 5–15 transverse bands and a total of less than 30 bands and dorsal spots; ventrals 128–146; lateral interspaces wider than lateral bars / *Natrix sipedon pleuralis* Cope, Midland Water Snake (p. 297)

25 a. Scale rows 29 or more at midbody; usually 4 prefrontals / *Pituophis melanoleucus melanoleucus* (Daudin), Northern Pine Snake (p. 259)

b. Scale rows less than 29; 2 prefrontals 26

26 a. Lower labials less than 8 (*Virginia valeriae*, Earth Snakes) 3

b. Lower labials 8 or more (Genus *Thamnophis*) 27

27 a. Lateral stripe on 2nd and 3rd scale rows / *Thamnophis sirtalis sirtalis* (Linnaeus), Eastern Garter Snake (p. 278)

b. Lateral stripe on 3rd and 4th scale rows 28

27b

28 a. Parietal spots usually fused, bright, and conspicuous; brown pigment of sides usually not extending onto ventral scutes, but if so, then covering less than 2/5 of the area of each scute / *Thamnophis proximus proximus* (Say), Western Ribbon Snake (p. 276)

b. Parietal spots often lacking; if present, then small, rarely fused or bright; brown pigment always extending onto ventral scutes and usually covering 2/5 or more of the area of each scute / *Thamnophis sauritus sauritus* (Linnaeus), Eastern Ribbon Snake (p. 274)

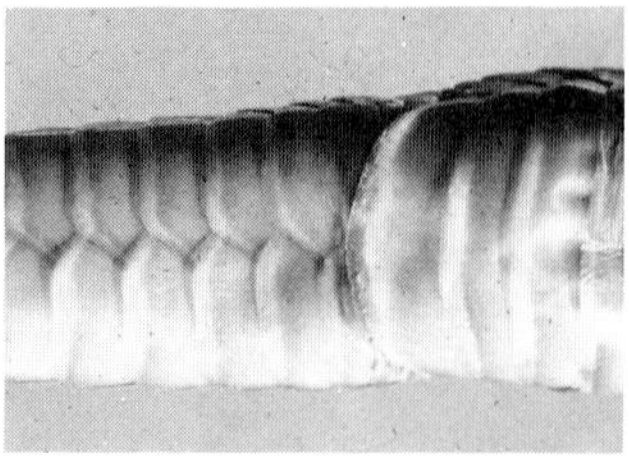

29a

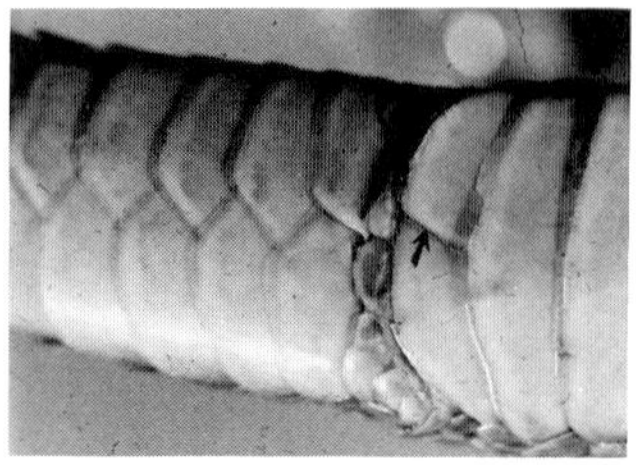

29b

29 a. Anal plate entire 30

b. Anal plate divided 36

30 a. Ventral surface light, without dark markings / *Cemophora coccinea copei* Jan, Scarlet Snake (p. 270)

b. Ventral surface with dark markings (Genus *Lampropeltis*) 31

31 a. Dorsal color black with a small white or yellow spot on numerous scales (*Lampropeltis getulus*, Kingsnake) 32

b. Dorsal pattern with red or with black-edged reddish or brownish blotches 33

32 a. With a white or yellow spot on many fewer than half the dorsal scales / *Lampropeltis getulus niger* (Yarrow), Black Kingsnake (p. 263)

b. With a white or yellow spot on many more than half the dorsal scales / *Lampropeltis getulus holbrooki* Stejneger, Speckled Kingsnake (p. 263)

33 a. Pattern of black-edged brownish to reddish dorsal blotches, only narrowly in contact with the 5th scale row / *Lampropeltis calligaster calligaster* (Harlan), Prairie Kingsnake (p. 261)

b. Pattern of rings or crossbands; or if blotches of brown, gray, or reddish, then in broad contact with 5th scale row 34

34 a. Usually with one anterior temporal; red areas continuous across belly / *Lampropeltis doliata doliata* (Linnaeus), Scarlet Kingsnake (p. 266)

b. Usually with 2 anterior temporals; red areas not continuous across belly 35

35 a. Dorsal saddles usually more than 35; often 2 rows of alternating lateral spots; a dark band on the posterior part of the prefrontals; a black-bordered light band from eye to angle of mouth; usually a light Y-shaped spot on back of head / *Lampropeltis doliata triangulum* (Lacépède), Eastern Milksnake (p. 266)

b. Dorsal saddles usually less than 35; only one row of alternating lateral spots; head markings poorly, if at all, developed / *Lampropeltis doliata syspila* (Cope), Red Milksnake (p. 266)

36 a. Scale rows more than 19 at midbody 37

b. Scale rows less than 19 at midbody 38

37a

37b

37 a. Preoculars absent; belly black and red / *Farancia abacura reinwardti* Schlegel, Western Mud Snake (p. 241)

b. Preoculars present; belly various but not black and red (Genus *Elaphe*) 13

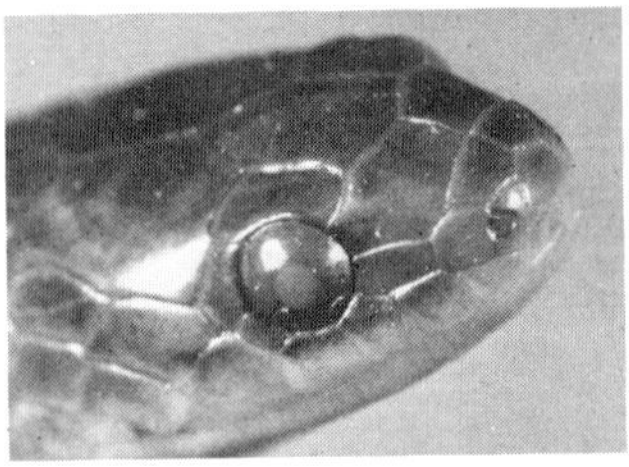

38a

38b

38 a. No preocular; loreal reaching eye 39

b. One or more preoculars present 41

39a

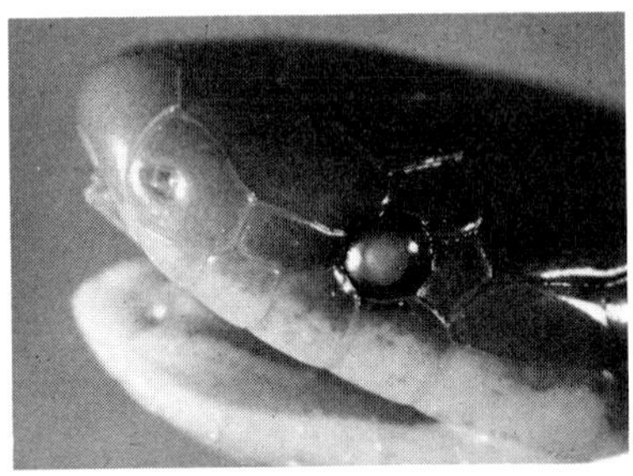

39b

39 a. Nasal plate divided; upper labials 6; scale rows 15–17 (*Virginia valeriae*, Earth Snakes) 3

b. Nasal plate undivided; upper labials 5; scale rows 13 (*Carphophis amoenus*, Worm Snake) 40

40 a. Internasals and prefrontals united into a pair of large shields / *Carphophis amoenus helenae* (Kennicott), Midwest Worm Snake (p. 239)

b. Internasals and prefrontals separate / *Carphophis amoenus amoenus* (Say), Eastern Worm Snake (p. 239)

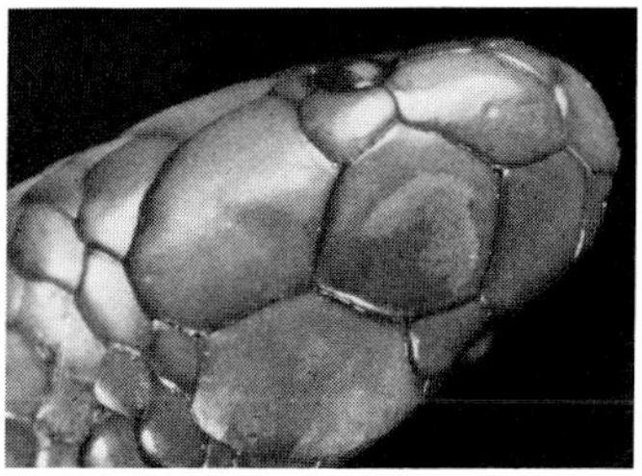

40a

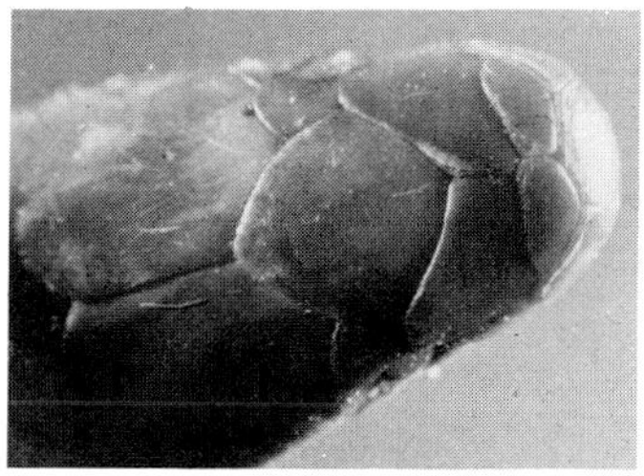

40b

41 a. Anterior temporals 2–3; a small preocular wedged between labials 42

b. A single anterior temporal; no small preocular wedged between labials 45

41a(1)

41a(2)

42 a. Scale rows 13 or fewer just anterior to vent; adults brown or back anteriorly; becoming lighter posteriorly / *Masticophis flagellum flagellum* (Shaw), Eastern Coachwhip (p. 253)

b. Scale rows 15 just anterior to vent (*Coluber constrictor*, Black Racer) 43

43 a. Back with 5 to 80 blotches, becoming obscured posteriorly / Juvenile *Coluber constrictor* Linnaeus, Black Racer (p. 250)

b. Back unmarked, uniform black or blue black 44

44 a. Basal hemipenial spine less than 3 times the length of its predecessor in the same row / *Coluber constrictor constrictor* Linnaeus, Northern Black Racer (p. 250)

b. Basal hemipenial spine 3 or more times the length of its predecessor in the same row / *Coluber constrictor priapus* Dunn and Wood, Southern Black Racer (p. 250)

45 a. Neck with a yellow ring just behind the head; belly yellow; 2 preoculars (*Diadophis punctatus*, Ringneck Snake) 46

b. Neck without a yellow ring; belly pink or red; one preocular / *Tantilla coronata coronata* Baird and Girard, Southeastern Crowned Snake (p. 272)

46 a. Belly usually unspotted but sometimes with a row of small imperfectly developed black spots in midline / *Diadophis punctatus edwardsi* (Merrem), Eastern Ringneck Snake (p. 243)

b. Belly conspicuously spotted with black, the spots showing a tendency to fuse into a single row or irregularly massed / *Diadophis punctatus stictogenys* Cope, Mississippi Ringneck Snake (p. 243)

Worm Snake

Carphophis amoenus (Say)

Recognition: A small, plain, chestnut or grayish, small-headed, burrowing snake. Belly and 1 or 2 rows of scales on either side pinkish. Length usually 6 to 10 inches, to 13. Scales perfectly smooth, opalescent, anal divided. Scale rows 13 at midbody. The young are usually darker dorsally than adults, and they have a more sharply bicolored appearance.

Variation: Two subspecies of worm snakes occur in Kentucky.

Carphophis amoenus amoenus (Say). Eastern Worm Snake. This subspecies is recognizable by the separation of the internasals and prefrontals, resulting in 4 plates between the frontal and rostral plates. Ranges throughout the Eastern Coal Field; meets and intergrades in a broad belt along the western edge of the area with:

Carphophis amoenus helenae (Kennicott). Midwest Worm Snake. The internasals and prefrontals are united on each side, resulting in only 2 plates between the frontal and rostral plates. Ranges throughout western and central Kentucky to near the Eastern Coal Field, where it meets and intergrades with the preceding.

Males have proportionally longer tails than females, and the difference is sufficiently great to be readily recognized by an experienced observer. Generally, the tail in males averages about 19 percent of the total length; in females, less than 15 percent.

Confusing Species: Similar to the brown snakes, which have strongly keeled scales. Most like earth snakes, which have weakly keeled scales in 15 or more rows, and the belly color does not extend up to include a full row of scales or more.

Habitat: Normally worm snakes remain under cover; logs, stumps, rocks, leaf mold, sandy soil—all supply abundant hiding

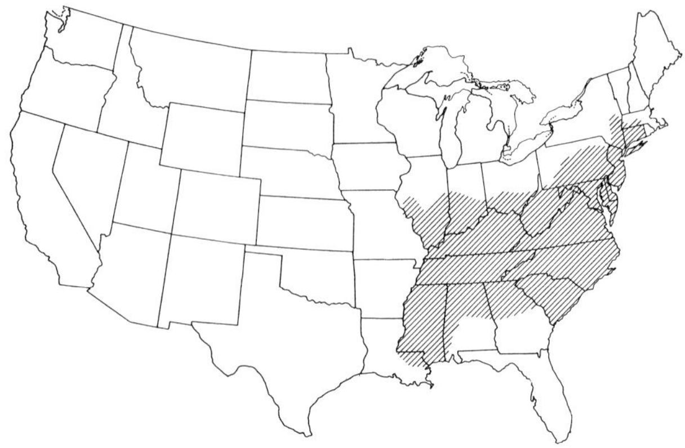

Range of the Worm Snake

places. They seem to be limited to wooded areas or to open fields bordered by woods and are more abundant in less cultivated areas.

Kentucky Distribution: Worm snakes occur throughout Kentucky and are probably the most abundant reptiles in the state. They are particularly abundant in the Eastern Coal Field, perhaps least common in the Blue Grass. Even there, however, they occur in considerable numbers along the cliffs bordering the Kentucky River.

Life History: A few successive warm days in March will bring worm snakes out of hibernation to absorb the heat beneath flat stones on open hillsides. On chilly days or nights, however, they go back into the earth, even well up into May.

A particular worm snake ranges over about a quarter acre and seems to know where all the good hiding places are; at least, it returns repeatedly to the same ones. An individual may remain under a rock, in summer, for upwards of 3 weeks without once coming out. Prolonged periods of inactivity seem common in this species.

From 2 to 5 eggs constitute a clutch, and they are usually laid about the middle of June. The individual eggs are surprisingly large for so small a snake; they average nearly an inch long and are between ¼ and ⅓ inch in diameter. They hatch in September or October, and the youngsters at hatching are about 4 inches long.

Worm snakes seem very selective in their diet; examination of the stomach contents of 116 specimens from Kentucky revealed only earthworms.

Remarks: Even when first handled, the little worm snake makes no effort to bite, although it will endeavor to make its way between the fingers with both head and tail. They make gentle pets and require little care other than a moist place to live, water to drink, and earthworms to eat.

Mud Snake XLIV

Farancia abacura (Holbrook)

Recognition: A large, shiny, iridescent, dark blue or black swamp snake with a boldly patterned red and black belly. The young are like adults in color and pattern. The eyes are degenerate, and the tail tip ends in a sharp spur. Length usually 36 to 54 inches, to 80. Scales smooth; anal plate divided. Scale rows 19 or 20 at midbody.

Variation: No geographic variation is recognized in Kentucky; *Farancia abacura reinwardti* Schlegel, the western mud snake, is the only subspecies occurring in the state. Males have an enlarged tail.

Confusing Species: The combination of black, unmarked dorsum and red and black belly is distinctive; no other Kentucky snake possesses these characteristics.

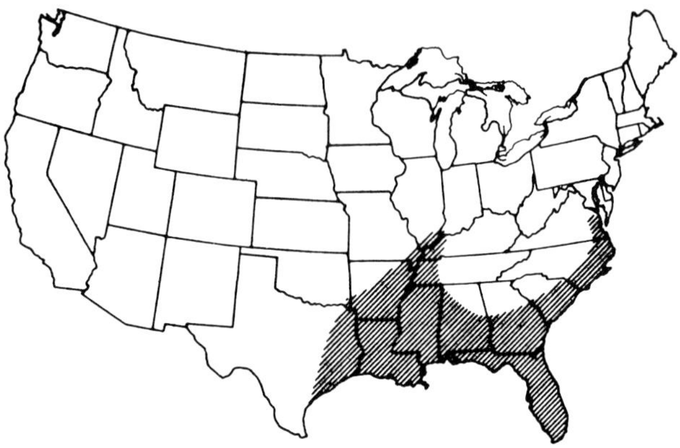

Range of the Mud Snake

Habitat: Swamps and muddy lowlands. Shallow sloughs with many rotten, water-soaked logs and much emergent vegetation provide apparently ideal habitats.

Kentucky Distribution: Throughout the Jackson Purchase in suitable habitats. The eastern limits are unknown, but suitable habitats are found at least as far east as Butler County, and it may well occur that far east in the state.

Life History: Mud snakes emerge from hibernation in late March or early April. Though it is not clear where they hibernate, there is evidence that they leave the swamps for holes on nearby higher ground.

Breeding habits of the Kentucky population are little known, but they lay eggs, numbering up to about 50, in cavities in the earth or in rotten logs. The individual eggs are creamy white, leathery shelled, and from 13/16 to 15/16 x 1⅝ to 1⅞ inches. Apparently the female remains coiled with the eggs until they hatch, absenting herself only to feed. The eggs hatch from mid-August to late September or October, and the hatchlings are from 6¼ to 9 inches long.

Mud snakes seem to feed almost exclusively on the large, aquatic salamanders *Siren* and *Amphiuma,* although an occasional frog is also taken.

Remarks: This completely harmless snake is locally considered to be highly venomous and, as the story goes, injects its venom by a stinger in the tail. When handled, a newly caught mud snake will make exploratory pricks with the spine on the end of its tail, and a large snake can sometimes barely break the skin, but the effect is more startling than harmful. Doubtless this habit has given rise to the "stinging snake" myth. Moreover, when at rest on a mud bank, the snake often lies in a rough loop; this probably accounts for the common belief that this is a "hoop snake," able to take its venomous tail in its mouth and roll like a hoop, to launch itself, tail-first, with deadly effect at an intruder.

The plain truth is that no snake can travel in such a manner, nor has any snake a sting in its tail, much less a venomous one.

Ringneck Snake XLV

Diadophis punctatus (Linnaeus)

Recognition: A relatively small, plain, dark, burrowing snake with a yellow collar and a yellow belly. Dorsal color variable, blue-black to slate gray, sometimes brownish. The young are darker above than adults. Neck ring complete dorsally or sometimes interrupted in the midline. Belly often plain yellow, sometimes with scattered dark spots forming a single row down the midline, sometimes with two rows of irregular dark spots down the belly. Length usually 10 to 16 inches, to about 23. Scales smooth; anal plate divided. Scale rows usually 15 (sometimes 16 or 17) at midbody.

Variation: Two subspecies occur in Kentucky.

Diadophis punctatus edwardsi. (Merrem). Northern Ring-

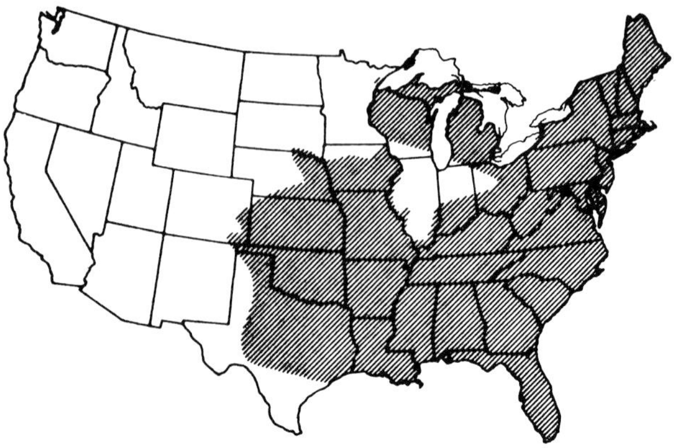

Range of the Ringneck Snake in the United States

neck Snake. Belly color plain yellow or with scattered dark spots sometimes aggregated in a single row down the midline. Chin usually without dark spots. Neck ring broad, rarely interrupted in the midline. Upper labials usually 8 on each side. Statewide in distribution except in the Jackson Purchase where it intergrades with and is replaced by:

Diadophis punctatus stictogenys Cope. Mississippi Ringneck Snake. Belly yellow, usually with a double row of irregular dark spots down the middle, but sometimes single, sometimes in attached pairs. Chin usually spotted with dark. Neck ring narrow, often interrupted in the midline. Upper labials usually 7 on each side. Occurs in the Jackson Purchase; meets and intergrades with the preceding eastward.

Males have fewer ventrals, more caudals, and proportionally longer tails than females.

Confusing Species: Juvenile brown snakes are our only other snakes with a neck ring, but they have keeled scales.

Habitat: A woodland species, rarely found more than a few yards from the woods. Quite secretive, they rarely emerge from hiding and are most often found beneath rocks, logs, stumps, or slabs, and sometimes beneath the loose bark of dead trees. A favored hiding place in spring is beneath a large, flat stone on the sunny, steep slope of a road cut in a wooded area.

Kentucky Distribution: Statewide; common, but most numerous in the Eastern Coal Field.

Life History: Ringnecks emerge from hibernation in late March or early April and remain active at least until October. They apparently mate soon after emergence.

The females are sexually mature when about 12½ inches long; they lay onc clutch of cggs per year. The number of eggs per clutch varies from 2 to 6, with 3 the average; larger females lay more eggs. On Big Black Mountain, in Harlan County, eggs are laid in late June and early July. The eggs are white, leathery shelled, and elongate (roughly 5/16 x 1 inch) and are laid in a moist site such as a rotten log or stump or in loose soil. They hatch in August or September, and the youngsters at hatching are about 5 inches long and look like miniature adults.

They eat almost any animal they can capture, subdue, and swallow—worms, insects and smaller vertebrates such as salamanders, frogs, lizards, and occasionally other snakes.

Remarks: These handsome little snakes are quite docile, rarely attempting to bite, even when first handled. If one does bite, it cannot even break the skin.

They are often mistakenly thought to be the young of the black racer.

Hognose Snake

Heterodon platyrhinos Latreille

Recognition: A medium-sized, stout-bodied snake with a pointed, keeled, and turned-up rostrum. When disturbed it flattens the head and body, hisses loudly, and strikes blindly. Undoubtedly the hognose snake is the most variable snake in Kentucky, both in color and pattern. The ground color may be tan, gray, red, or black. The markings, usually in the form of a series of dorsal blotches alternating with lateral spots, may be brown, gray, or black. Some individuals are almost black with a faint pattern, and others are completely black with no indication of a pattern at all. Such black individuals may be encountered anywhere in the state, but seem more abundant than the patterned individuals in eastern Kentucky. The young have larger heads with usually more contrasting patterns than adults. Length usually 16 to 30 inches, to 43. Scales keeled; anal plate divided. Scale rows at midbody 23 to 27, usually 25.

Variation: The variation in color and pattern seems not clearly correlated with geography; *Heterodon platyrhinos platyrhinos* Latreille, the eastern hognose snake, is the only subspecies known in the state. Males have fewer ventrals, more caudals, and longer tails than females.

Confusing Species: Sometimes confused with the copperhead. The latter, among other differences, always has a facial pit between the eye and nostril which the hognose never has.

Habitat: Varied, from wooded hillsides to open cultivated fields with a preference for rather open, sandy areas. In Kentucky it is most often encountered along woodland edges.

Kentucky Distribution: Statewide; nowhere abundant but common in most areas; perhaps least common in the Blue Grass.

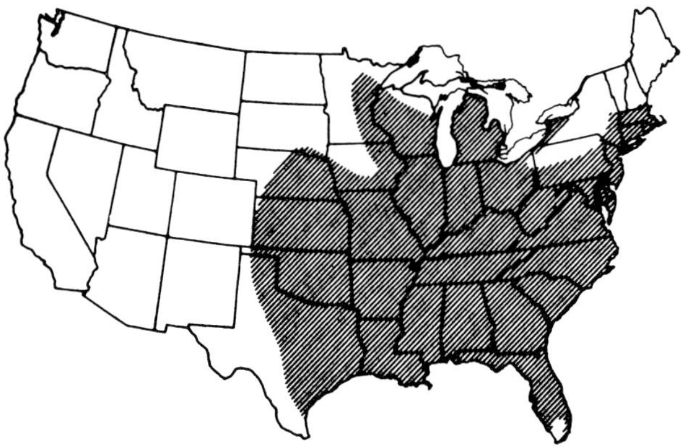

Range of the Hognose Snake in the United States

Life History: Hognose snakes emerge from hibernation in late March or early April and remain active until late October or early November. Mating occurs shortly after emergence, and the eggs are laid in June or July in a clutch of up to 35, though usually 15 to 20, deposited in some sheltered spot such as underneath a flat stone. The eggs, elongate and thin shelled, range from ½ to 1 inch x 1 to 1½ inches, usually about ¾ x 1¼ inches; the size is apparently correlated with the size of the female. The eggs hatch in August or sometimes in late July or early September, and the newly hatched babies are from 6½ to about 8½ inches long.

Food consists primarily of toads, but fish, salamanders, frogs, lizards, snakes, birds, mice, and chipmunks, as well as assorted worms and insects, are known to have been eaten.

Remarks: When approached, the hognose will flatten its head to double its width, puff up its body, hiss loudly, all the while slowly waving the head and front portion of the body from side to side. Occasionally the animal will strike viciously, though it usually misses, and in any event makes no attempt to bite. If these actions fail to drive the intruder away, the snake frequently will roll over and "play dead." After writhing, the snake rests on its back with the jaws widely agape and the tongue lolling out. If placed on its belly, it immediately rolls over on its back; this reaction is the only one that can readily be elicited. In spite of its fearsome behavior, the hognose snake is quite harmless. When handled, however, it smears the handler with evil-smelling fluids.

Rough Green Snake XLVI

Opheodrys aestivus (Linnaeus)

Recognition: A long, slender, grass-green snake with a yellowish-green belly. Newly hatched youngsters are gray-green until they first shed, a few days after hatching. They have larger heads and longer tails than adults. Length usually 24 to 36 inches, to 42. Scales keeled; anal plate divided. Scale rows 17 at midbody.

Variation: No geographic variation has been observed in Kentucky; *Opheodrys aestivus aestivus* (Linnaeus), the eastern rough green snake, is the only subspecies occurring in the state. Males on the average have longer tails, more caudals, and fewer ventrals than females, but because of variations, the sexes cannot reliably be separated by external features.

Confusing Species: The only plain green snake known to occur in Kentucky. The smooth green snake, *Opheodrys vernalis* (Harlan), possibly occurs in Kentucky, although it is not now known to do so. The smooth green snake resembles the eastern rough green snake but is less elongate and has smooth scales with no indication of keels. It should be looked for especially in extreme eastern Kentucky and along the Ohio River.

Habitat: Generally woods and woodland edges, swampy or dry. Usually they are found on the ground, probably because they are more conspicuous there and hence more often seen. They spend a great deal of time in bushes and low trees and while motionless are difficult to detect. They frequently lie along a branch or across a cluster of leaves with the head and front part of the body extended several inches beyond the support. A green snake in such a pose will often "freeze" and remain until approached within a few inches before gliding gracefully away over the vegetation.

Kentucky Distribution: Throughout Kentucky, almost surely occurring in every county; least abundant in the Inner Blue Grass but even there common along the bluffs of the Kentucky River.

Life History: Green snakes usually do not emerge before late April or early May. The females lay up to 12 or so eggs (usually from 4 to 10) about mid-July. The eggs are elongate, smooth, and shiny, with a rather thick, tough, leathery, white shell. They average nearly ½ inch in diameter and about 1 inch in length. They hatch usually in August or early September, and the newly hatched young are about 7 or 8 inches long.

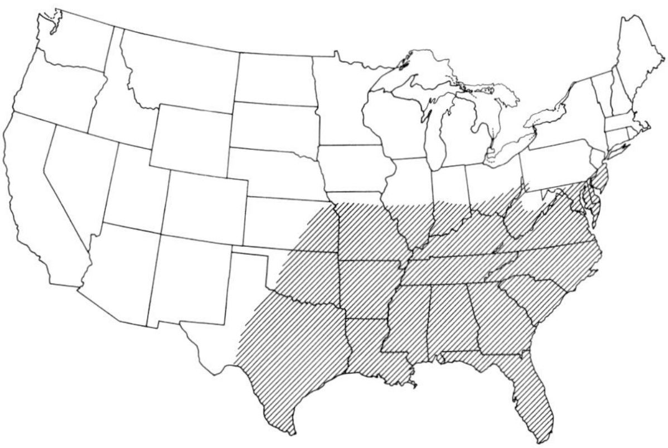

Range of the Rough Green Snake in the United States

Food consists almost exclusively of insects, with a few spiders, snails, and an occasional tree toad.

Remarks: One never ceases to marvel at the stories people tell of snakes. Consider what is told of the harmless and nearly friendly little green snake. According to the story, they are exceedingly venomous and bite on the slightest provocation. The victim always turns green and dies of a laughing fit. The truth is that the rough green snake is not only nonvenomous, but ordinarily refuses to bite.

Black Racer XLVII

Coluber constrictor Linnaeus

Recognition: A long, moderately slender, satiny, black snake, with a dark belly and white chin, throat, and neck. The young are strikingly different from adults in appearance. They possess a series of 50 to 80 reddish dorsal blotches that become progressively more indistinct and crowded posteriorly. The tail is usually gray and the belly speckled with brown. The sides of the head are spotted with white, and there is a small preocular wedged between two of the upper labials. The juvenile pattern is gradually lost, the ventral spotting disappearing last. Most racers over 30 inches in length show few or no traces of the juvenile pattern. Length usually 30 to 60 inches, to 73. Scales smooth, anal plate divided. Scale rows 17 at midbody.

Variation: Two subspecies, externally nearly indistinguishable, are recognized in Kentucky.

Coluber constrictor constrictor Linnaeus. Northern Black Racer. With proximal spines of hemipenes not excessively enlarged, less than 3 times the length of their predecessors in the same row. Iris usually brown. Ranges throughout eastern Kentucky; western limits poorly known, but it meets and intergrades in west-central Kentucky with:

Coluber constrictor priapus Dunn and Wood. Southern Black Racer. With the proximal spines of the hemipenes much enlarged, 3 or more times the length of their predecessors in the same row. Iris often red or orange. Ranges throughout the Jackson Purchase and up the Ohio River. Eastern limits of the range poorly known, but it meets and intergrades with the preceding in west-central Kentucky.

Males cannot reliably be recognized from females by external characteristics

Confusing Species: Mud snakes have red and black bellies; melanistic garter snakes have keeled scales. Most like large dark rat snakes, which have keeled mid-dorsal scales with the body shaped like a loaf of bread in cross section. The black phase of the coachwhip is quite similar but is usually lighter on the tail, reddish on the sides of the tail, and has 13 rather than 15 scale rows just anterior to the base of the tail.

Habitat: Deep woods to cultivated fields. They are quite arboreal and often flee upward when closely pursued, but they will also take refuge in burrows in the earth. In Kentucky they

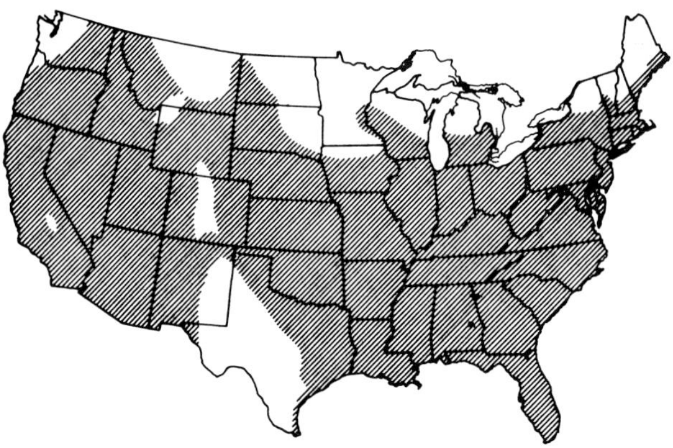

Range of the Black Racer in the United States

are most often encountered along woodland edges or overgrown fence rows.

Kentucky Distribution: Statewide; common, but least common in the Inner Blue Grass.

Life History: Black racers usually emerge from hibernation about the first of April and mate in May. Egg laying occurs usually in June, and the average clutch of 20 to 25 eggs is deposited in a cavity under a stone, in a rotten log, or buried in loose earth or humus. The individual egg is white and oval, with a rough, thick, leathery shell, and about ⅝ to 1 inch x 1 to 1¾ inches in size. They often become lumpy or irregular and discolored by dirt during incubation. They usually hatch from mid-July to mid-August, and the newly hatched youngsters are from 11 to 13½ inches long.

Food consists of a variety of animals. In spring small mammals constitute a large part of the diet, but insects make up most of the food in late summer. Other prey includes frogs and toads, snakes, lizards, small turtles, birds, and bird eggs.

Remarks: There are some apparently valid reports of black racers deliberately attacking persons in Kentucky. In one such case, a large racer approached a man and wrapped itself about his lower leg before it was killed. The attack occurred in May and was likely performed by a particularly aggressive male. When cornered, the racer often vibrates the tail tip rapidly, creating a rattling noise in dry leaves. If pressed, they will strike viciously and accurately. They do not make particularly satisfactory pets, as they remain vicious and often do not feed well.

Coachwhip

Masticophis flagellum (Shaw)

Recognition: A very large, lithe, slender, dark snake that becomes lighter in color toward the tail. The color ranges from dark brown or black at the head to light brown at the tail with considerable variation on individual snakes; some are quite dark throughout their length. The scalation of the long, slender tail suggests a braided whip, hence the name. The belly color grades off like that of the back, from dark to light. The young are light brown with many closely spaced, transverse, dark dorsal bands which become less distinct posteriorly. Length usually 36 to 60 inches, to about 102. Scales smooth; anal plate divided. Scale rows 17 at midbody.

Variation: No data are available on geographic variation in Kentucky; *Masticophis flagellum flagellum* (Shaw), the eastern coachwhip, is likely the only subspecies occurring in the state.

Confusing Species: A normally colored individual should not be confused with any other Kentucky snake. Dark individuals are similar to black racers, which have 15 scale rows just anterior to the tail, while coachwhips have 13. Might be confused with an old, dark rat snake, which has keeled scales in the mid-dorsal region.

Habitat: Dry uplands most commonly. It may be found in woodlands, open fields, and on dry, rocky hillsides. It often hides in a burrow when pursued.

Kentucky Distribution: Unknown. A small population of this species occurred a few years ago in Barren and Edmonson counties, but the colony may now be extinct. Since it still occurs in southern Illinois, eastern Missouri, and southwestern Tennessee, there is a possibility that it occurs in other parts of western Kentucky.

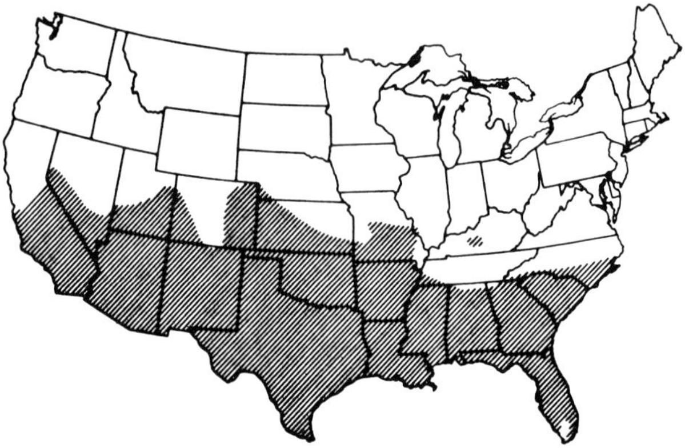

Range of the Coachwhip in the United States

Life History: No life history data are available on the coachwhip in Kentucky. Farther west they mate in April or May, and the eggs are laid in June or July in clutches of about 12. The individual eggs are much like black racer eggs, but larger, averaging about ⅞ x 1½ inches but sometimes as large as 1 3/16 x 2 inches. The eggs hatch in August or September, and the newly hatched young are from 12 to 16 inches long.

Food consists mainly of small mammals, young birds, snakes, and lizards.

Corn Snake XLIX

Elaphe guttata (Linnaeus)

Recognition: A large, slender-bodied, beautifully marked orange or red snake. The dorsal pattern consists of a series of 40 or fewer red, reddish-brown, or orange blotches narrowly edged with black, arranged in a single row down the mid-dorsal region. They are bordered on each side by a series of similar alternating spots. The ground color is orange, brown, or gray. There is a prominent, spear-shaped mark on the head and neck, created by 2 extensions of the first dorsal blotch that form a dark point between the eyes. Belly white or yellowish, strongly marked with a black checkerboard pattern; underside of tail is striped. In cross section, the body is shaped like a loaf of bread. The young are more contrastingly marked, with darker blotches, and have relatively larger heads than adults. Length usually 30 to 48 inches, to 72. Scales weakly keeled; anal plate divided. Scale rows 27 (rarely 29) at midbody.

Variation: No geographic variation is recognized in Kentucky; *Elaphe guttata guttata* (Linnaeus), the corn snake, is the only subspecies occurring in the state. Males have more caudal scutes and relatively longer tails than females.

Confusing Species: Similar to the milk snakes, which have smooth scales, a single anal plate, and no striping under the tail. The rat snakes are darker in color and lack the spear point between the eyes.

Habitat: From open woodlands to cultivated fields. It is most often found along woodland edges and overgrown fence rows

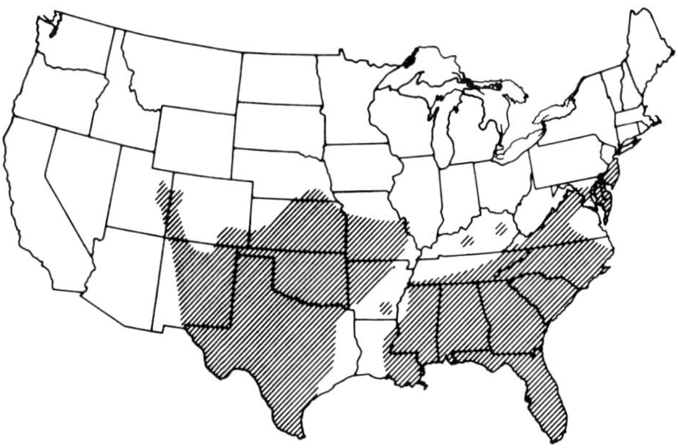

Range of the Corn Snake in the United States

or about farmsteads and spends considerable time in burrows or beneath such shelters as stones, logs, and boards.

Kentucky Distribution: Poorly known. Apparently the distribution is not continuous in Kentucky, for the species has been recorded from one locality in eastern Kentucky and from the Mammoth Cave area in Barren and Edmonson counties. It is nowhere common, although an occasional individual is encountered, especially in or near Mammoth Cave National Park.

Life History: The corn snake emerges from hibernation in late March or April, but comes up to lie briefly in the sun on warm days earlier in the season. It is oviparous, laying a clutch of about 15 to 18 eggs, but no data are available on its breeding habits in Kentucky.

Food consists largely of small mammals which are killed by constriction. Birds or bird eggs are sometimes eaten.

Rat Snake L

Elaphe obsoleta (Say)

Recognition: This is one of the two longest snakes in Kentucky, also one of the most variable in color and pattern. The young are strongly patterned dorsally with dark gray or brown blotches on a pale grayish ground color. There is a dark postocular stripe that extends from the eye to the corner of the mouth where it stops abruptly. With increasing age, the ground color becomes darker and the dorsal pattern, more obscure; however, many individuals retain the juvenile pattern throughout life. In others, by the time adulthood is reached, the back is a shiny black, with few if any traces of a pattern left except in the narrow spaces between the scales; these show only when the animal is distended. The belly is marked with gray or brown on a yellowish or white background. The chin and throat are plain white or sometimes yellowish. In cross section the body is shaped like a loaf of bread. Length usually 30 to 72 inches, to 102. Scales weakly keeled; anal plate divided. Scale rows 23 to 27 at midbody.

Variation: Extremely variable, with both age and geography. Two subspecies are recognized in Kentucky.

Elaphe obsoleta obsoleta (Say). Black Rat Snake. This subspecies becomes increasingly darker with age, and most individuals over 30 inches retain few traces of the juvenile pattern. Old adults are quite black. Ranges across Kentucky to near the Jackson Purchase where it meets and intergrades with:

Elaphe obsoleta spiloides (Duméril, Bibron, and Duméril). Gray Rat Snake. This form retains the juvenile pattern throughout life but becomes somewhat darker with age. However, a pattern of dark blotches on a lighter background is nearly always obvious. Usually the dark spots of this snake are brown on a gray ground color. Ranges throughout the Jackson Purchase eastward through the Land Between the Lakes to meet and intergrade with the preceding in the western portions of the Mississippian Plateau and Western Coal Field.

Males have more caudal scutes and slightly longer tails than females.

Confusing Species: Similar to the corn snake but lacks the spear point between the eyes. Black racers and coachwhips have smooth scales, with bodies rounded in cross section and with a small preocular wedged between the upper labials. Milk snakes and kingsnakes have smooth scales and single anal plates.

Habitat: Varied, ranging from dense woodlands to suburban homes. It is normally an inhabitant of woods and fields and is an excellent climber, sometimes seeking shelter in a hollow high in a tree. It frequently enters barns and other farm buildings and occasionally finds its way into residences.

Kentucky Distribution: Statewide; common. It is probably the most abundant large snake in most of Kentucky, par-

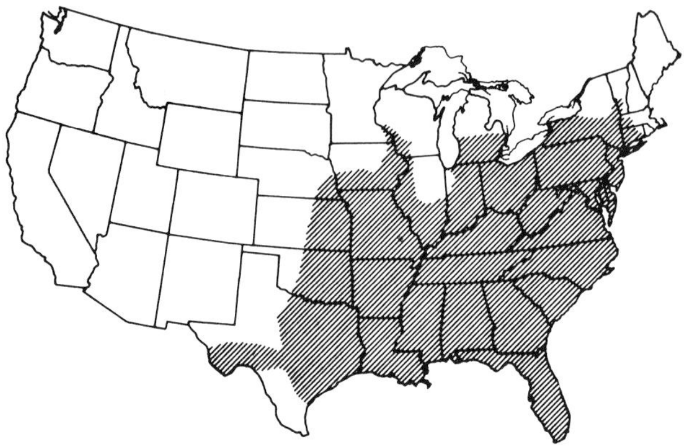

Range of the Rat Snake in the United States

ticularly common in the Inner Blue Grass and the Jackson Purchase.

Life History: Rat snakes usually emerge from hibernation in April and mate in May or June. The eggs, in clutches of 6 to 24, are usually laid during July, in sheltered spots, often buried in loose earth or humus. They are white, elongate, with a thick leathery shell usually measuring about 1 x 1½ inches but varying with the size of the female. They hatch usually in late August or September, and the newly hatched youngsters are about 14 inches long.

They feed on a large variety of small mammals, from baby mice to adult gray squirrels and half-grown rabbits. Many bird eggs and birds, especially nestlings, are also eaten. They are highly efficient mouse-catchers about farm buildings, although large individuals sometimes enter henhouses and swallow chicken eggs. Though obviously preferring warm-blooded prey, a hungry rat snake will not disdain frogs, salamanders, lizards, and other snakes.

Remarks: In eastern Kentucky the common name of "pilot black snake" is often applied to this species. According to many residents, the snake is supposed to lead copperheads and rattlesnakes away from danger, thus acting as their "pilot." The belief has no basis in fact. In central Kentucky, the snake is thought to milk cows, hence the common name of "cowsucker." Again, there is no basis for the belief.

Pine Snake LI

Pituophis melanoleucus (Daudin)

Recognition: A large, moderately stout-bodied, black and white or yellowish snake with a small, pointed head and a loud hiss. The dorsal pattern consists of a mid-dorsal series of large black blotches that often become more brown posteriorly. Ground color white, yellowish, or light gray. The young are patterned like the adults, but the ground color is paler, often with a reddish tinge. This species has 4 prefrontals, rather than the 2 of our other colubrids. Length usually 36 to 60 inches, to 83. Scales keeled; anal plate single. Scale rows 29 to 33 at midbody.

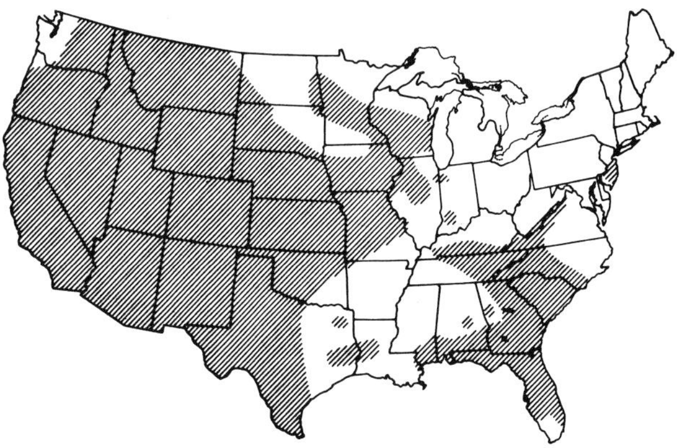

Range of the Pine Snake in the United States

Variation: There are insufficient specimens from Kentucky to determine if geographic variation exists across the state; *Pituophis melanoleucus melanoleucus* (Daudin), the northern pine snake, is the only subspecies known to occur in the state. Males have more caudals, fewer ventrals, and relatively longer tails than females.

Confusing Species: Might be confused with the gray rat snake, which has a divided anal plate and only 2 prefrontals.

Habitat: Principally dry, sandy, piney woods; also other areas, particularly along dry, wooded ridges. Although it can and does sometimes climb trees, the pine snake is basically a burrowing animal.

Kentucky Distribution: Incompletely known. It occurs in the area of Kentucky Lake in western Kentucky and the vicinity of Mammoth Cave National Park. There are also records from Whitley and Harlan counties in southeastern Kentucky. It probably can be found completely across southern Kentucky, from Harlan County to Calloway County and northward in west central Kentucky at least to Hart and Edmonson counties.

Life History: When disturbed, the pine snake throws itself into a coil, vibrates the tail rapidly, and strikes repeatedly and viciously, all the while uttering a series of hisses that can be heard up to 100 yards away. Although the display is largely bluff, some individuals will bite viciously and repeatedly if sufficiently provoked. In captivity it usually soon becomes quite docile.

Our lack of knowledge of the pine snake is doubtless due in part to its secretive habits. It seems to spend much time underground and is apparently active at night. Probably the pine snake is more common than the few records indicate; it is simply not seen.

Little is known of the breeding habits of the animal in Kentucky. From 7 to 12 eggs are laid about mid-July in a burrow some 6 to 8 inches deep. The individual egg is elongate and variable in size, from 1 to 1 13/16 x 1 13/16 to 2⅝ inches; larger snakes usually lay larger eggs. When freshly laid, the eggs have a thin, tough, parchmentlike shell and adhere to one another. They hatch in August or September, and the newly hatched young are about 14 to 16 inches long.

Food consists mostly of small mammals and an occasional bird or bird egg. The snake is a powerful constrictor and kills its prey in this manner.

Remarks: Any accurate observations on the distribution or life history of this species in Kentucky would be valuable.

Prairie Kingsnake LI

Lampropeltis calligaster (Harlan)

Recognition: A rather large, moderately slender, blotched or dark snake with a white belly checkerboarded with brown squarish blotches. There are two color phases of this snake, a blotched and a dark one. The blotched form has about 60 brown, reddish, or sometimes greenish, black-edged markings, with 2 rows of similar but smaller alternating blotches on each side. Ground color brownish gray to tan. The animal darkens with age until some old adults are often almost uniformly colored; however, this general darkening is frequently complicated by the development of 4 faint, dusky, parallel, longitudinal stripes. The young are more contrastingly marked and are more reddish than adults. Length usually 24 to 40 inches, to 50. Scales smooth, anal plate single. Scale rows usually 25 (23–27) at midbody.

Variation: No geographic variation is recognized in Kentucky; *Lampropeltis calligaster calligaster* (Harlan), the prairie king-

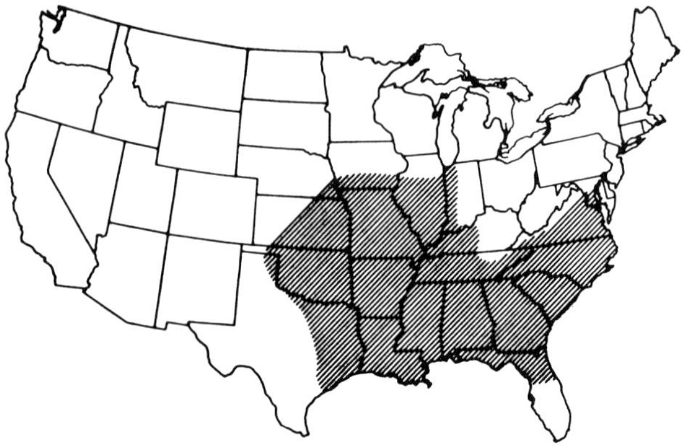

Range of the Prairie Kingsnake

snake, is the only subspecies occurring in the state. Males average more caudal scutes than females, but the sexes cannot reliably be separated by external features.

Confusing Species: Most like the milk snake, which has the belly marked with black and 21 scale rows at midbody. Similar to some rat snakes, which have weakly keeled scales and a divided anal plate. Well-marked individuals are similar to the corn snake, which has faintly keeled scales and a divided anal plate.

Habitat: A prairie species, usually inhabiting old fields and grasslands. It is found on occasion, however, in open woodlands or in cultivated fields.

Kentucky Distribution: Throughout the Jackson Purchase eastward across the Western Coal Field and onto the Mississippian Plateau. The eastern limits of its distribution are not clear, but it seems absent east of a north-south line through Barren and Meade counties.

Life History: Few data are available on this species in Kentucky. They usually emerge from hibernation on the first warm days in March but retire again during cold spells. They apparently mate soon after emergence, and the female deposits her 5 to 14 eggs in late June or early July in cavities in the earth. The eggs are elongate, white, and about ¾ x 1¾ inches. They hatch in August or early September after an incubation period of about 1½ months. The newly hatched young are about 8 to 10 inches long.

Small mammals constitute the major part of the diet, but amphibians, reptiles, birds, and insects are also eaten.

Kingsnake LII

Lampropeltis getulus (Linnaeus)

Recognition: A large, glossy, black, smooth-scaled snake with a tiny white or yellowish dot on many of the dorsal scales. In most individuals the dots are widely scattered over the back and sides; in some, the dots are arranged in a series of poorly defined crossbands; in others, the dots are essentially confined to the sides of the body. Belly checkered with black and yellowish, more yellowish anteriorly but decreasing posteriorly with a consequent increase in dark pigment. Length usually 36 to 48 inches, to 56. Scales smooth; anal plate single. The young are more obviously patterned than adults; juveniles of the 2 races cannot reliably be separated. Scale rows 21 to 23 at midbody.

Variation: There is considerable variation in number and arrangement of the light spots on the dorsum, individually, with age, and geographically. Two subspecies are recognized in Kentucky.

Lampropeltis getulus niger (Yarrow). Black Kingsnake. Usually with only a few light spots arranged to form a faint chainlike dorsal pattern, or frequently with the pattern completely obscured and only a few scattered light spots more abundant on the sides than on the back. This is the common kingsnake of Kentucky, ranging across the state into the Jackson Purchase where it meets and intergrades with:

Lampropeltis getulus holbrooki Stejneger. Speckled Kingsnake. Typically a "salt and pepper" snake with a great profusion of light dots scattered randomly over the back and sides. Nearly every scale bears a light dot. The kingsnakes of the extreme western and southern sections of the Jackson Purchase, although not completely typical of this race, closely approach it and are here considered members of the southern subspecies. This race meets and intergrades with the preceding in the Jackson Purchase.

Males average more caudals and longer tails than females, but the overlap is such that the sexes cannot reliably be separated by external features.

Confusing Species: Most like the black racer, which lacks light spots on the dorsal scales, has a divided anal plate, and has a small preocular wedged between two upper labials. Similar to old, dark black rat snakes, which have weakly keeled scales and a divided anal plate. Might be confused with some of the large, dark water snakes, but all these have strongly keeled scales.

Habitat: From dry, open woods to weedy pastures. It seems most often encountered along roadsides, overgrown fence rows, and woodland edges. In those areas it may often be found under logs, stones, or other sheltering objects.

Kentucky Distribution: There is some question about the distribution of this species in Kentucky. It is widespread and probably occurs throughout the state. However, it is rare in the Blue Grass and is perhaps absent from at least the northern half of the region. It seems most common in the Eastern Coal Field.

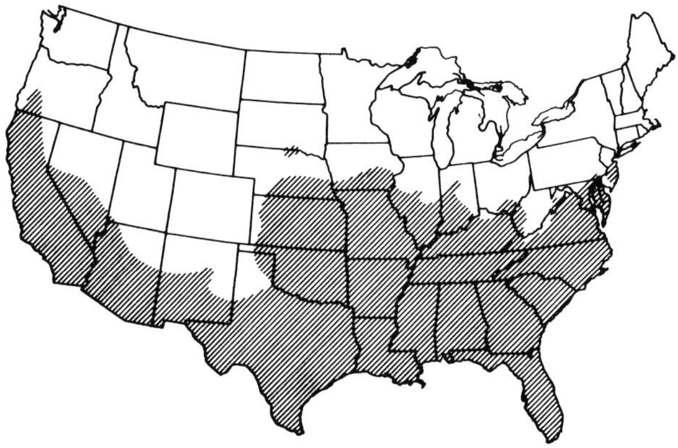

Range of the Kingsnake in the United States

Life History: Kingsnakes emerge from hibernation in late March or early April and remain active at least until the middle of October. Apparently mating occurs soon after emergence, and the female lays her eggs in a cluster of about 12 in June or early July. The creamy white, elongate eggs average about ¾ to 1 x 1¼ to 1¾ inches in size. The size is variable with larger females laying larger eggs. The eggs hatch in about 2½ months, and the youngsters, which measure about 9 to 12 inches in length, are abroad usually by late August or early September.

The kingsnake commonly feeds on other snakes and seems to possess an immunity to the venom of both rattlesnakes and copperheads. Although it will eat rattlers and copperheads (assuming, of course, that it is big enough to subdue and swallow them), it does not deliberately seek them. It will take any mammal, bird, or snake it can, even including a smaller kingsnake.

It is a constrictor and kills its prey by coiling about it and squeezing it to death. The first kingsnake I ever saw I discovered many years ago when as a curious youngster I investigated a series of squeaks and squeals coming from a clump of grass. I found a black kingsnake coiled about a meadow mouse, crushing it to death. The mouse's miseries were soon ended, and it became a meal for the snake as I watched in wonderment.

Remarks: Stories of fights between kingsnakes and rattlesnakes or copperheads are frequently told in Kentucky; I have heard at least a dozen, and some of them were wonderful. In many of them the kingsnake, every time it was bitten, crawled away to feed on the leaves of a small plant nearly always conveniently growing nearby. After a few bites the kingsnake completely recovered from the effects of the venom and recommenced the fight. In most of the stories the observer pulled up the plant and hid it; without the plant the kingsnake, when next bitten, "just coiled up and died."

Like many snake stories, this one has some basis in fact (kingsnakes are apparently immune to the venom of our venomous snakes), but the greater part of the story is pure fabrication.

Milk Snake LIII

Lampropeltis doliata (Linnaeus)

Recognition: This species is represented in Kentucky by three races, one of which so unlike the other two that it is treated separately here and below.

1) Milk Snake (two subspecies). A rather slender, moderately large snake with a series of large, red, reddish-brown, or brownish, black-bordered blotches down the middle of the back. Usually there is a lateral row (sometimes 2) of similar but smaller blotches on each side, alternating with the dorsal row. Ground color gray or tan. Frequently there is a V or Y-shaped light patch on the nape. Belly boldly but irregularly checkerboarded with black on a white background. Length usually 20

to 36 inches, to 48. Scales smooth; anal plate single. Scale rows 21 to 23 at midbody.

2) Scarlet Kingsnake. A handsome snake, banded with brilliant red, black, and yellow bands that go completely around the body. The red bands are bordered on each side by black, and the front of the head is red. Length usually 12 to 20 inches, to about 30. Scales smooth; anal plate single. Scale rows 19 at midbody.

Variation: Three subspecies are recognized in Kentucky.

Lampropeltis doliata doliata (Linnaeus). Scarlet Kingsnake. Throughout most of its range the scarlet kingsnake is merely one of several subspecies constituting the milksnake *(Lampropeltis doliata)* complex. In Kentucky, however, the scarlet kingsnake occurs with the eastern milksnake and is apparently reproductively isolated from it. It behaves as a distinct species here. It is therefore treated separately in the ensuing paragraphs; the sections headed Confusing Species, Habitat, etc. following the next two subspecies apply only to those two subspecies and not to the present one.

The scarlet kingsnake is most like the scarlet snake, *Cemophora coccinea,* which has a plain white belly. Might be confused with one of the more prominently banded milksnakes, which have 21 or more scale rows at midbody and a checkered belly. Most often confused with the venomous coral snake, a southern species which does not occur in Kentucky. In the coral snake the snout is black, and the red bands are bordered by yellow. Remember the little poem to separate the two:

Red and yellow,
Kill a fellow.
Red and black,
Venom lack.

This snake prefers wooded areas and is occasionally dug from a rotten log or found beneath the bark of a dead tree. It is shy and secretive, normally emerging from hiding only at night or after a heavy rain, and is adept at worming its way into small cracks and crevices, either into logs or rocks, or to considerable depths in the ground. The first one I ever saw in Kentucky was discovered several feet underground by workmen blasting out a cut for a road in Floyd County.

The scarlet kingsnake is apparently quite widespread in Kentucky, although its range is but sketchily known. It occurs at least to Rowan County in the north, Floyd County in the east, McCreary and Whitley counties in the south, and Edmonson County in the west. These constitute the only records, and it is likely quite rare. It is so highly secretive, however, that its rarity may be more apparent than real.

Essentially no life history data are available on this snake in Kentucky. Farther south it emerges from hibernation in March or April and lays its few eggs in July. The newly hatched young are almost 5 inches long.

Its food includes lizards, snakes, mice, insects, and earthworms.

Any accurate observations on this little known snake in Kentucky would be a real contribution to our knowledge.

Lampropeltis doliata triangulum (Lacépède). Eastern Milk Snake. Usually with a light V or Y-shaped mark on the nape and 1 or 2 rows of small, dark-bordered, reddish spots on either side, alternating with the dorsal blotches. Usually no light neck ring. Ranges throughout the Eastern Coal Field, the Knobs, and the Blue Grass to the western segment of the Knobs, and westward in the south across the Mississippian Plateau to about the eastern edge of the Western Coal Field where it meets and intergrades in a broad band with:

Lampropeltis doliata syspila (Cope). Red Milk Snake. Usually with a light neck ring and no V or Y-shaped mark on the nape. Dorsal blotches large and extending well down onto the sides. Lateral blotches much reduced and small or virtually absent, especially toward the head. Occurs throughout western Kentucky, eastward to near the eastern limits of the Western Coal Field where it meets and integrades in a broad band with the preceding.

Confusing Species: Similar to the scarlet snake, *Cemophora coccinea*, which has a plain, white belly. Some are similar to the scarlet kingsnake, which has 19 scale rows at midbody with the color bands going completely around the body. Probably most often confused with the copperhead, which has a unicolored coppery head, facial pits, vertical pupils, fangs, and but a single row of caudal scales. Similar to the corn snake, which has weakly keeled scales and a divided anal plate.

Habitat: Varied, from river bottoms to high mountain slopes and from open fields to dense woodlands. Apparently feeding by night, milk snakes are secretive by day, usually spending those hours in some sheltered spot, beneath a stump, log, or stone. They commonly enter buildings and may take up residence in a building harboring a plentiful supply of mice. They are excellent climbers and often rest along the rafters or other elevated timbers in a building.

Kentucky Distribution: Statewide; common, particularly in the Blue Grass. Every county in Kentucky likely supports a population of milk snakes.

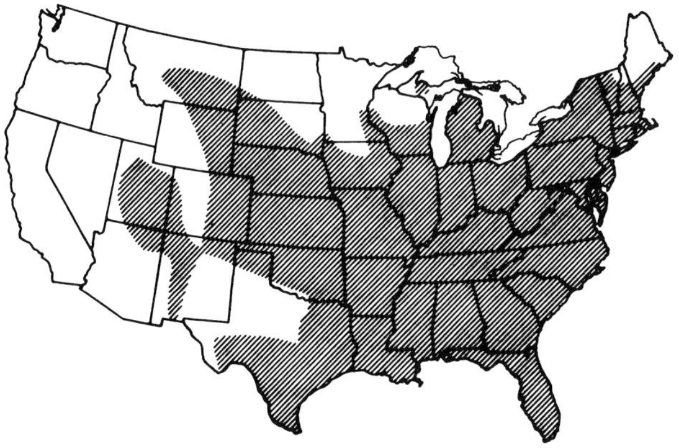

Range of *Lampropeltis doliata* in the United States

Life History: Milk snakes emerge from hibernation in late March or early April and remain active at least until mid-October. They mate in May and June, and the females lay their clutches of 6 to 24 eggs in late June and July. The eggs are deposited in loose earth, humus, old sawdust piles, or the like. The individual eggs are elongate, with a tough, white, leathery shell and are from ⅝ to 1 x 1¼ to 1½ inches in size. They hatch in about 6 to 8 weeks, and the young, some 6½ to 10 inches long, emerge in late August or September. The babies in their autumnal wanderings often enter residences.

Food consists almost exclusively of small mammals, which are killed by constriction. A few birds are eaten, and the milk snake is not averse to feeding on other snakes.

Remarks: Over much of its range the milk snake is reported to take milk from cows, but there is not one shred of valid evidence that it does.

Scarlet Snake LIV

Cemophora coccinea (Blumenbach)

Recognition: A medium-sized, slender snake that appears to be banded with red, black, and white or yellow rings; the snout is red, and the belly is pale and unmarked. The dorsal pattern consists of a series of red blotches extending well down onto the sides and surrounded by a black border, which is broadest in the mid-dorsal region. The ground color is often white but can vary from grayish to yellowish. From a dorsal view, the snake appears banded, but from a lateral view the true pattern becomes immediately apparent. Young individuals have dark flecking along the sides, which disappears with age, and the red saddles are pinkish. Very old individuals are often quite dull and dark. Length usually 12 to 24 inches, to about 33. Scales smooth; anal plate single. Scale rows usually 19 (17–20) at midbody.

Variation: No geographic variation is recognized in Kentucky; *Cemophora coccinea copei* Jan, the northern scarlet snake, is the only subspecies known in the state. Males have fewer ventrals, more caudals, a proportionally longer tail, and more body blotches than females; however, the overlap is such that the sexes cannot reliably be separated by external features.

Confusing Species: Similar to the scarlet kingsnake, in which the red, black, and yellow bands extend across the belly. Resembles some milk snakes, which have dark blotches on the belly.

Habitat: A woodland species, occurring in pine, hardwood, and mixed forests. The scarlet snake is a burrowing animal, usually found under the shelter of a log or stone, hidden in leaf litter, or burrowed into sandy soil. It sometimes emerges at night, to prowl about on the forest floor.

Kentucky Distribution: It is known to occur across the state south and west of a line connecting Jefferson and Bell counties, although there is a questionable record from Rowan County

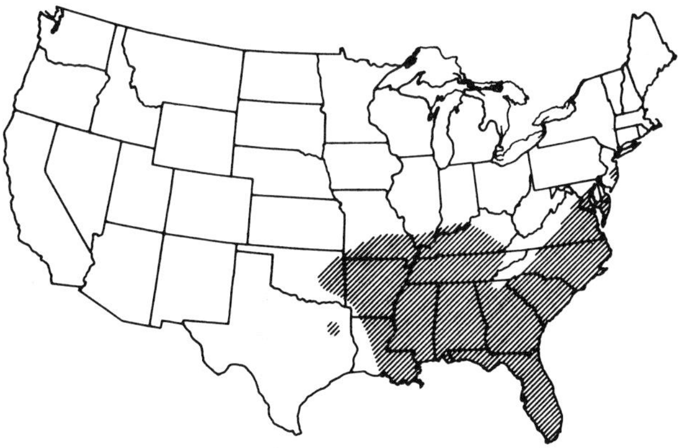

Range of the Scarlet Snake

in northeastern Kentucky. This handsome snake is rarely encountered; whether because of its secretive nature or its scarcity is unclear.

Life History: By early May the scarlet snake has emerged from hibernation. Although the time of mating is unknown, it is probably shortly after emergence. No data are available on breeding habits in Kentucky, and very little elsewhere. They apparently lay from 3 to 8 greatly elongate, leathery-shelled eggs (nearly 1½ inches long), presumably in late June or early July.

Although no detailed study of its food habits has been undertaken, largely because of the scarcity of specimens, enough data are in to indicate that small lizards and reptile eggs constitute the bulk of its diet. Insects, amphibians, and small mammals make up a substantial portion of the remainder of the diet.

Remarks: So little is known of the life history and distribution of this animal in Kentucky that every scrap of information about it should be recorded.

Crowned Snake LIV

Tantilla coronata Baird and Girard

Recognition: A tiny, brown or reddish-brown snake with a black headcap and a light band across the back of the head followed by a black band from 3 to 5 scales wide. The black band on the back of the head is suggestive of a tiny crown. The black on the head and nape extends down on each side to about the level of the mouth. The belly is usually white, but sometimes yellowish or pinkish. The young have darker heads and lighter bodies than adults. Length usually 8 to 10 inches, to 13. Scales smooth; anal plate divided. Scale rows 15 at midbody.

Variation: No geographic variation is known to occur in Kentucky; *Tantilla coronata coronata* Baird and Girard, the southeastern crowned snake, is the only subspecies occurring in the state. Apparently males and females cannot be separated by external features.

Confusing Species: Similar in size and behavior to earth, brown, and worm snakes, but the dark headcap, light collar, and dark "crown" are distinctive. Any small, black-headed snake discovered in or near the Jackson Purchase should be examined closely; it is possible that *Tantilla gracilis hallowelli* Cope, the northern flatheaded snake, occurs in that area, although it has not been recorded. It is similar to the southeastern crowned snake but usually has a paler headcap with a concave posterior margin, no collar or "crown," and a salmon pink belly.

Habitat: Dry, rocky, southern exposures in open woods or along woodland edges. The crowned snake is quite secretive, usually found beneath a stone, log, or the litter of the forest floor.

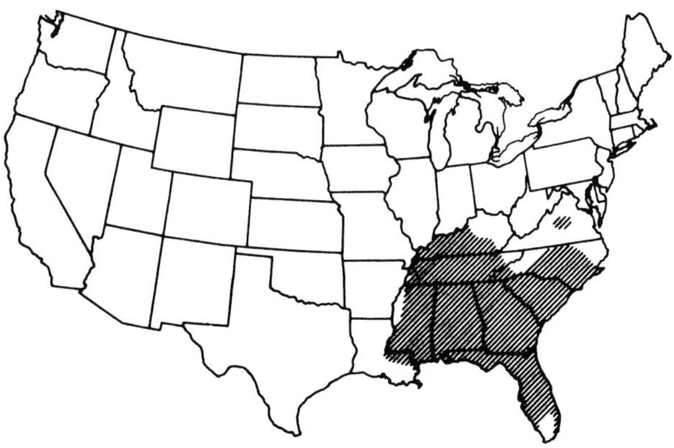

Range of the Crowned Snake

Kentucky Distribution: Spotty and poorly known. It occurs in Bullitt, Edmonson, and Hopkins counties and probably throughout the Mississippian Plateau and the Western Coal Field. Although it seems nowhere abundant, it is fairly common on some south-facing wooded slopes in Hopkins County.

Life History: Crowned snakes usually emerge in late March or early April and remain active until at least mid-October. In Georgia they mate in April and early May; the breeding pairs are often found beneath a scrap of bark at the base of a rotting stump. Usually 2 or 3 eggs (sometimes 1 or 4) are deposited in a cavity in a pile of humus or in a well-rotted log. The eggs are cylindrical, white, and about ¼ to 5/16 x 9/16 to 15/16 inch in size. They apparently hatch in August or September, and the newly hatched young are about 3 to 3½ inches long.

Food consists of various invertebrates. Centipedes, insects, and earthworms are known items of diet.

Remarks: This little snake has almost microscopic fangs in the rear of the jaws and minute poison glands. The mouth is so tiny, however, that the fangs cannot possibly be brought into play on a human.

Eastern Ribbon Snake LV

Thamnophis sauritus (Linnaeus)

Recognition: A medium-sized, long-tailed, slender, dark snake with three brilliant yellow stripes; one is mid-dorsal and 2 are lateral, on the 3rd and 4th scale rows. The dorsal stripe may sometimes be rather greenish or orange. A double row of vague black spots may appear between the stripes. The belly is plain yellowish or greenish in the middle, and the ends of

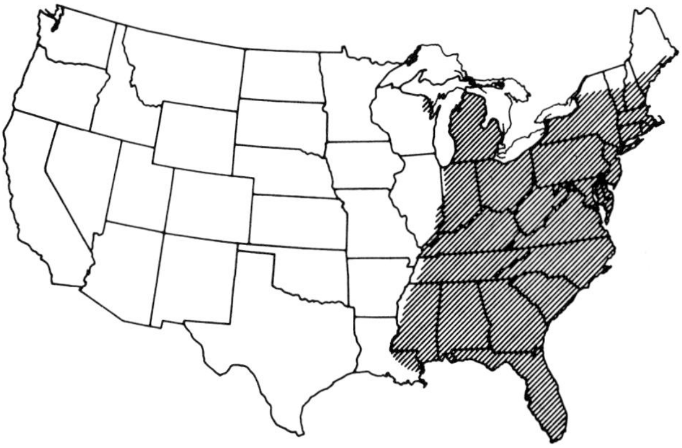

Range of the Eastern Ribbon Snake in the United States

the scutes are usually invaded by the brown lateral pigment. Parietal spots are usually lacking, but if present are small and usually paired. The young have relatively shorter tails and larger heads than adults and are brown dorsally; they darken with age to dark brown or black. Ordinarily, there are 7 upper labials. Length usually 18 to 30 inches, to 38. Scales keeled; anal plate single. Scale rows 19 at midbody.

Variation: Too few specimens are available from Kentucky to determine if geographic variation exists; *Thamnophis sauritus sauritus* (Linnaeus), the eastern ribbon snake, is the only subspecies known to occur in the state. Males average shorter than females, but have proportionally longer tails and more caudals.

Confusing Species: Most often confused with the eastern garter snake, which has a much shorter tail and less obvious stripes, with the lateral stripe on the 2nd and 3rd scale rows. Most like the western ribbon snake, which has the parietal spots large, bright, and usually fused, and with the brown color of the sides usually not extending onto the ends of the ventrals.

Habitat: Widely varied, but all near permanent or at least semipermanent bodies of water. They often climb into low vegetation, although rarely more than 4 feet off the ground.

Kentucky Distribution: Apparently statewide. Too few data are available for definite conclusions; the species is known only from Rowan County in the east and Hickman County in the west. It seems nowhere abundant and is quite rare or even absent in most areas.

Life History: Very little data are available on the life history of these snakes in Kentucky. They have emerged from hibernation by mid-April (likely in March) and mate shortly after emergence. The young are born in late July or early August. An average litter contains about 10 to 12 youngsters, but may run from 3 to 26. The youngsters range from about 5 to 7½ inches in length at birth.

The diet of the eastern ribbon snake consists chiefly of fishes and amphibians. Frogs and toads are eaten more often than salamanders, but this is perhaps a reflection of their availability.

Remarks: Any accurate observation on this species in Kentucky would be a worthwhile contribution.

Western Ribbon Snake LV

Thamnophis proximus (Say)

Recognition: A medium-sized, long-tailed, slender, dark snake with three brilliant stripes. The mid-dorsal stripe may be orange, yellow, or greenish; the lateral stripes, on the 3rd and 4th scale rows, are yellowish or greenish. Belly plain, greenish or yellowish, and the dark pigment of the sides barely, if at all, invades the ends of the ventrals. The parietal spots are fused, bright, and large. Usually there are 8 upper labials. The young are more slender, shorter tailed, and larger headed than adults. Length usually 18 to 30 inches, to 44. Scales keeled; anal plate single. Scale rows 19 at midbody.

Variation: No geographic variation is recognized in Kentucky; *Thamnophis proximus proximus* (Say), the western ribbon snake, is the only subspecies known to occur in the state. Males are slightly smaller than females and have relatively longer tails and more caudals.

Confusing Species: Most often confused with the eastern garter snake, which has the lateral stripes on the 2nd and 3rd scale rows. Most like the eastern ribbon snake, which has separate, small, dark parietal spots, and a broad invasion of the ends of the ventral scutes by the lateral color.

Habitat: Closely confined to the vicinity of permanent or semipermanent water, either standing or running. It is found most frequently in open, grassy, or weedy areas but sometimes enters wet woodlands. Individuals will sometimes climb several feet into weeds or low bushes.

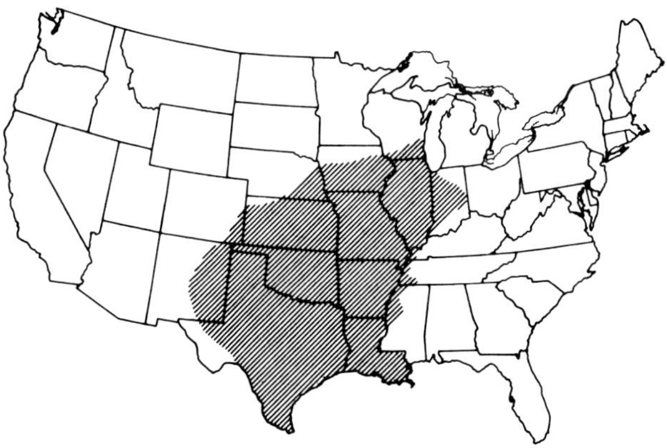

Range of the Western Ribbon Snake in the United States

Kentucky Distribution: Known only from near the north end of Reelfoot Lake in Fulton County. It likely occurs in other areas of the Jackson Purchase, especially in the lowlands along the Mississippi River.

Life History: Little data are available on the life history of this species in Kentucky. They apparently hibernate in rocky or gravelly banks near the water and emerge with the first warm days of spring. They mate soon after emergence, and the young, usually about 12, are born in late July or August. At birth the youngsters are usually between 6 and 9 inches long.

Food is chiefly confined to fishes and amphibians. Frogs, toads, and tadpoles are eaten more often than salamanders. These snakes seem to forage as actively by night as by day when the temperature permits.

Remarks: Any accurate observation on the distribution, ecology, or natural history of this species would be valuable.

Garter Snake — LVI

Thamnophis sirtalis (Linnaeus)

Recognition: A medium-sized, slender, dark snake with three more or less obvious stripes, a mid-dorsal one and a lateral one on the 2nd and 3rd scale rows on each side. Ground color ranges from almost black through browns and reds to blue-green; the stripes are usually yellowish but may be brown, green, or bluish. Usually there is a double row of alternating spots between the stripes, and they may be evident or indistinct. In some individuals the stripes are predominant; in others, the spots are more obvious. The belly is greenish or yellowish, with 2 rows of indistinct dark spots partially hidden by the overlapping ventrals. The young usually have brighter, more contrasting colors and relatively larger heads than adults.

Length usually 16 to 30 inches, to 48. Scale keeled; anal plate single. Scale rows 19 at midbody.

Variation: No geographic variation is recognized across Kentucky; *Thamnophis sirtalis sirtalis* (Linnaeus), the eastern garter snake, is the only subspecies occurring in the state. There is a great deal of individual variation, as indicated above, but it seems not to be correlated with geography. Males have longer tails and more caudal scutes than females.

Confusing Species: Most like the ribbon snakes, which have the lateral stripes on the 3rd and 4th scale rows. The queen water snake has a strongly patterned belly and a divided anal plate.

Habitat: Widely varied, from deep woodlands through brushy or grassy pastures to city lots. It seems to prefer areas where there is moisture such as the vicinity of streams, swamps, drainage ditches, or seepage areas.

Kentucky Distribution: Statewide; abundant. This is one of the commonest and probably the most widely known harmless snake in Kentucky; it is still abundant, but the population is steadily declining.

Life History: Garter snakes spend the winter buried in the mud at the bottom of drainage ditches and other such wet areas and emerge with the first warm days of spring. They are sometimes out in late February and often emerge in numbers in early March. However, one cannot depend on seeing them in numbers much before late March or early April. Vegetation is low when they emerge, and they are frequently observed. They are no less common in summer and far more abundant in the fall, when the young are about. However, they are less frequently seen at those times because of the greater amount of cover.

Mating usually takes place on the first warm days after

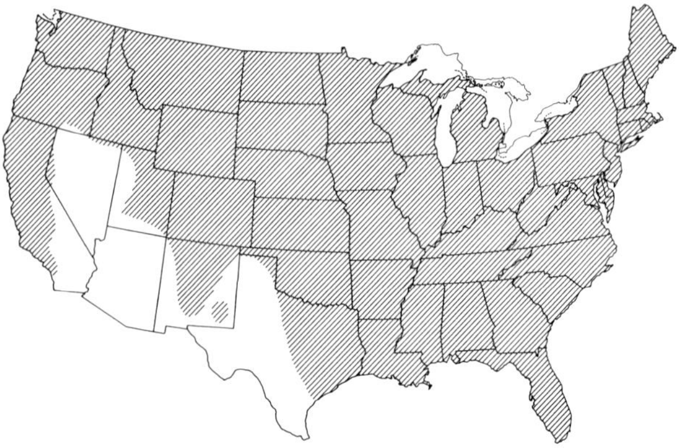

Range of the Garter Snake in the United States

emergence, and the young are born in late July and early August. The youngsters are from 5 to 9 inches long at birth, and about 15 to 20 is an average brood. One mother produced 85 youngsters at one time but this is quite exceptional. There is no parental care, and as soon as the babies are born, they and their mother go their separate ways. By late October most garter snakes have gone into hibernation, but occasionally an individual may be encountered abroad in November.

Food consists of a wide variety of animals with earthworms predominant. Amphibians are commonly eaten, depending on availability. Garter snakes in the woods and along small streams commonly feed on salamanders; those in more open areas eat more frogs and toads. Insects are eaten, particularly by the young, but apparently are not an important food item. Occasionally a small bird or mammal may be eaten.

Remarks: The garter snake is a timid creature and will flee when approached. If cornered, it will often flatten the body, displaying the color pattern prominently, and present an apparently fierce appearance. If further molested, it will excrete a vile-smelling fluid from glands near the base of the tail. Some individuals will bite savagely, but the bites are of no more consequence than scratches. When handled, most individuals become completely docile in a few moments and make interesting pets. They feed well in captivity and usually take readily to raw chopped fish.

The myth that snakes swallow their young to protect them and regurgitate them later when danger is passed may have originated with the garter snake. An overly imaginative person, seeing a handful of young snakes emerging from the body of a snake he had just mangled, might well later imagine that he remembered seeing the mother swallow all those youngsters. Neither this nor any other snake swallows its young as a protective measure.

Earth Snake LVI

Virginia valeriae Baird and Girard

Recognition: A small gray or reddish brown snake without conspicuous markings. Sometimes has tiny scattered, dark flecks on the dorsum, especially on the head. Belly white or yellowish. Loreal scale present, horizontal, and reaching the eye. Length usually 6 to 10 inches, to about 12¾. Scales smooth or weakly keeled; anal plate divided. Scale rows 15 or 17 at midbody.

Variation: Two subspecies are recognized in Kentucky.

Virginia valeriae valeriae Baird and Girard. Eastern Earth Snake. Scales mostly smooth but faintly keeled posteriorly. Dorsal coloration gray or brownish gray, often with tiny, scattered black spots. Scale rows 15 at midbody. Distribution poorly known, but it occurs at least throughout the Eastern Coal Field and the eastern Knobs. It seems absent from the Blue Grass, and in the southern Knobs and the eastern third of the Mississippian Plateau likely meets and intergrades with:

Virginia valeriae elegans Kennicott. Western Earth Snake. Scales mostly faintly keeled. Dorsal coloration grayish-brown to reddish. Scale rows 15 anteriorly, 17 at midbody. Ranges throughout western Kentucky, eastward at least into Nelson County. Meets and intergrades with the preceding race as indicated above.

Males have fewer ventrals, more caudals, and proportionally longer tails than females.

Confusing Species: Most like the worm snakes, which have perfectly smooth scales in only 13 rows, small heads, and tiny eyes. Similar to the brown snakes, which have strongly keeled scales and no loreal.

Habitat: A woodland species which is highly secretive and seldom encountered, apparently spending most of the time buried in the leaf mold of the forest floor but sometimes found beneath stones and logs.

Kentucky Distribuiion: Poorly known, but probably statewide. However, it seems quite rare, if in fact not completely absent from the Blue Grass. It seems nowhere common in Kentucky,

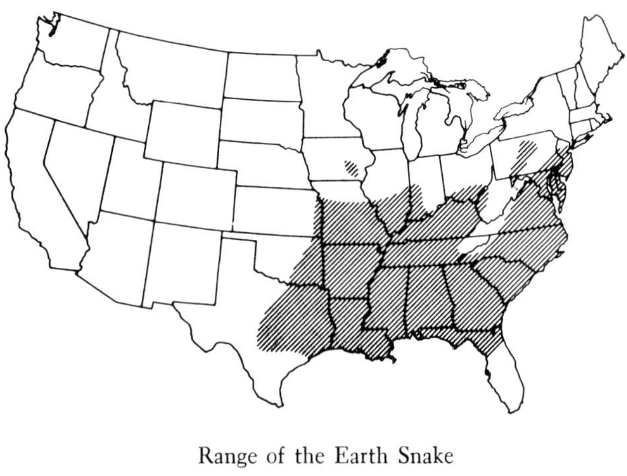

Range of the Earth Snake

but this scarcity may be more apparent than real because of its highly secretive nature and small size. Several individuals have been found in Bell County, and scattered individuals elsewhere across the state.

Life History: Little is known of this species in Kentucky. They are active by April 1 and remain active at least through September. Time of mating is unknown but probably is in early spring. From 4 to 10 (sometimes as many as 18) tiny young, usually about 4 inches long, are born in late July or August.

Food consists of assorted worms, slugs, snails, and insects and other arthropods found in the leaf litter of the forest floor.

Brown Snake LVII

Storeria dekayi (Holbrook)

Recognition: A small brown or gray snake with 2 parallel rows of spots down the back. Usually the mid-dorsal area, about 4 scales wide, is lighter than the sides, and the spots on one side may couple with their partners on the other side by a narrow dark line across this light area, creating a ladderlike effect. There is a second row of spots along the sides, usually alternating with the more dorsal row, but these are sometimes quite indistinct. There is usually a dark streak ranging downward on the side of the head behind the eye. The young have a distinctive white or light gray collar just behind the head. The belly is pale, brownish, pinkish, or yellowish, and unmarked except for small black spots toward the ends of the ventral scutes. Length usually 8 to 14 inches, to about 18. Scales keeled; anal plate divided. Scale rows 17 at midbody.

Variation: Quite variable across the state; 2 subspecies are recognized.

Storeria dekayi dekayi (Holbrook). Northern Brown Snake.

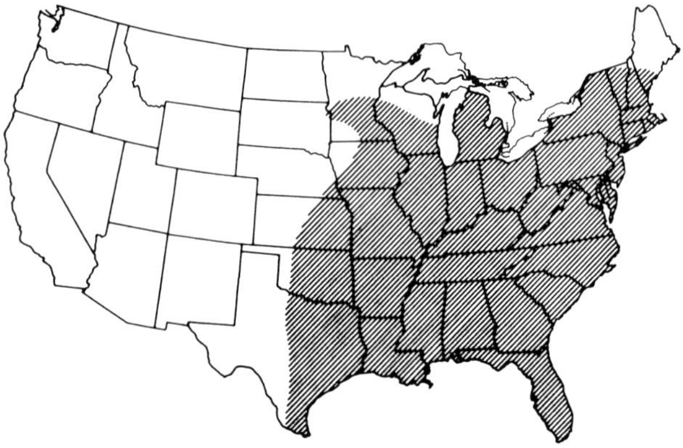

Range of the Brown Snake in the United States

Light mid-dorsal area rarely crossed by dark lines connecting the dorsal spots on opposite sides. Sum of ventrals and caudals 175 or fewer. Ranges through the Eastern Coal Field, westward through the Blue Grass and the Knobs; western limits of the range poorly known, but it intergrades broadly with:

Storeria dekayi wrightorum Trapido. Midland Brown Snake. Most of the paired dorsal spots connected by narrow dark lines across the mid-dorsal light area. Sums of ventrals and caudals 176 or more. Ranges throughout the Jackson Purchase and the Land Between the Lakes. Eastern limits of the distribution are poorly known, but it intergrades with the preceding where they meet.

Males have fewer ventrals, more caudals, and longer tails than females.

Confusing Species: Often confused with spotted garter snakes, which have an undivided anal plate and usually stripes along the sides. Earth snakes have a long, horizontal loreal, while the brown snakes have a vertical preocular. Ringneck snakes are easily confused with young brown snakes, but they have smooth scales. Red-bellied snakes usually have a red belly, 3 light spots on the occiput, and only 15 scale rows at midbody.

Habitat: Widely varied, from upland woods through grassy or brushy pastures and urban lots to lowland marshes and swamps. It is seldom encountered, spending much of the time hiding beneath a rock, log, slab, or other cover. It is sometimes seen crawling about in a grassy or weedy area and is almost invariably mistaken for a young garter snake. In the swamplands of western Kentucky it sometimes crawls 2 or 3 feet up on vegetation in spring and basks in the sunlight.

Kentucky Distribution: Statewide. The species seems rare in southeastern and central Kentucky, but is common in the northeastern section and particularly so in the Jackson Purchase. Because of its usually secretive habits it may not be so scarce as it seems.

Life History: This species usually appears in early April. Mating occurs soon thereafter, but there is evidence of a fall mating as well. The young, measuring about 3¼ to 4½ inches at birth and numbering up to 24, are usually born in late July or August. Brown snakes go into hibernation, sometimes in aggregations of several dozen, in late October or early November.

Food is various, but earthworms seem the favored diet. However, small treefrogs, newly transformed toads, insects, snails, slugs, and tiny fish are sometimes eaten.

Red-bellied Snake LVII

Storeria occipitomaculata (Storer)

Recognition: A small, rather plain snake, usually with a red belly and a series of 3 light spots on the occiput. The young have more distinct occipital spots than adults. Sometimes the belly color is orange or yellowish or blue-black in some exceptional cases. On occasion the 3 spots on the occiput fuse together to form a light bar across the back of the head. The

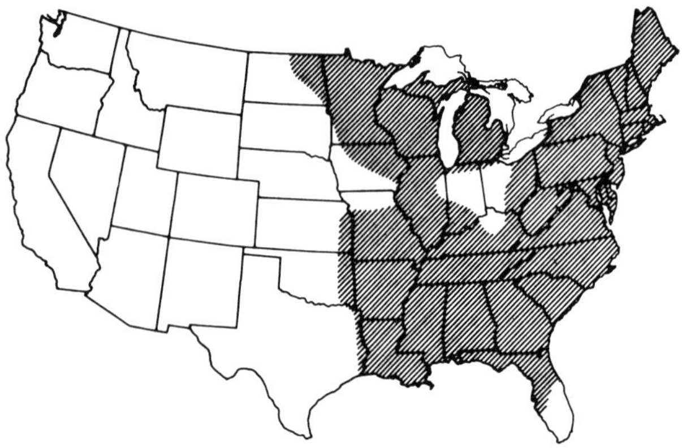

Range of the Red-bellied Snake in the United States

dorsal color is usually grayish or brownish but sometimes red-brown or tan. There are faint indications of 4 narrow dark stripes and sometimes a broad mid-dorsal light stripe; on rare individuals all the stripes, both light and dark, are present. Occasionally the dorsal color is almost black. Regardless of color, the animal is rather easily identified—keeled scales; anal plate divided; no loreal; the preoculars in contact with the posterior part of the nasal; and 15 scale rows at midbody. Length usually 6 to 10 inches, to 16.

Variation: No geographic variation is recognized across Kentucky; individual variation is indicated above. *Storeria occipitomaculata occipitomaculata* (Storer), the northern red-bellied snake, is the only subspecies occurring in the state. Males have more caudals, longer tails, and fewer ventrals than females.

Confusing Species: Earth snakes (*Virginia*) have smooth or only faintly keeled scales; Kirtland's water snake has 2 rows of dark spots down the belly.

Habitat: A woodland species, seldom venturing out into meadows or pastures. It is quite secretive, usually found hiding beneath a flat stone or board. Sometimes, however, one is seen climbing about on low vegetation much like the closely related brown snake.

Kentucky Distribution: Apparently absent from both the Inner and Outer Blue Grass; otherwise, occurring across the state, though nowhere abundant.

Life History: These small snakes usually emerge from hibernation in early April. They apparently soon mate, but there is evidence that a late summer or early fall mating sometimes occurs. The tiny young, from about 2 to 4 inches in length and numbering from 4 to 10, are born in late July, August, or early September.

Food consists largely of earthworms, slugs, and insects. A variety of other invertebrates are eaten and occasionally small salamanders.

Queen Water Snake LVIII

Regina septemvittata (Say)

Recognition: A medium-sized, slender, brown or grayish water snake with a yellowish stripe along each side on the 2nd scale row and the top half of the 1st and with three narrow dark stripes on the back, which become less apparent with age and are not discernible in many individuals. Belly yellowish with 4 dark stripes, one on each end of the ventral scutes and one on either side of the midline. The belly stripes are usually more obvious in the neck region. The young are like adults but with more contrasting stripes. Length usually 16 to 30 inches, to about 36. Scales keeled; anal plate divided. Scale rows 19 at midbody.

Variation: No geographic variation is recognized in Kentucky; *Regina septemvittata septemvittata* (Say), the queen water snake, is the only subspecies occurring in the state. Males have longer tails and more caudals than females. The caudals average about 80 in males, about 72 in females; usually this characteristic will serve to separate the sexes.

Confusing Species: May be confused with garter snakes, which have no stripes on the belly and a single, undivided anal plate. Most likely to be confused with Graham's water snake, *Regina grahami* Baird and Girard, which has not been recorded from Kentucky but may occur in the Jackson Purchase. It differs from the queen snake in having a broader yellow band on the lower side (on scale rows 1, 2, and 3) and a plain yellow belly, sometimes with a dark line or a single row of spots down the midline.

Habitat: This species is basically an inhabitant of small rocky streams in wooded areas, but is not confined there. In the Blue Grass it is abundant in rocky streams in open pastures, and it has been found in swampy woods.

Kentucky Distribution: The distribution of this species in Kentucky is poorly known. Published range maps show it

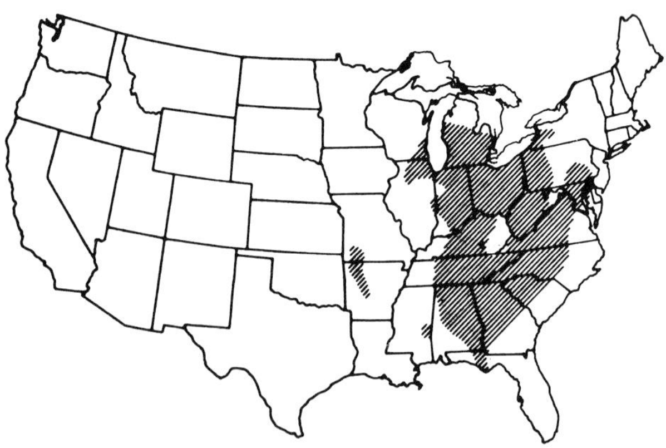

Range of the Queen Water Snake in the United States

occurring across roughly the eastern three-fourths of the state; however, it seems completely absent in the Eastern Coal Field except perhaps in the most northeastern section. It is abundant in the Blue Grass and occurs commonly in the Knobs. South of the Knobs, it ranges across the eastern end of the Mississippian Plateau, westward to at least Cumberland County.

Life History: This is usually one of the early snakes to appear in spring; often they emerge in March but sometimes not until April. They mate soon after emergence.

The young, numbering up to about 20 (usually 10 or 12) per litter, are born between late July and early September under the shelter of a flat stone at the edge of a stream. At birth the youngsters are usually between 7 and 9 inches in length.

In autumn these snakes sometimes gather in considerable numbers about a preferred hibernation site. Although on occasion an individual may be seen in November, hibernation usually begins by late October.

This species feeds almost exclusively on crayfish, but they will also eat minnows, frogs, salamanders, and snails.

Remarks: Most species of water snakes are evil-tempered and will bite viciously if captured, but not the queen water snake. Of the hundreds I have handled, not one has even attempted to bite, no matter where held.

Green Water Snake LVIII

Natrix cyclopion (Duméril, Bibron, and Duméril)

Recognition: A large, heavy-bodied, dark water snake without distinctive dorsal markings. It may be readily separated from all our other water snakes by the presence of a short row of suboculars interposed between the eye and the upper labials. Belly dark, with a series of light spots or half-moons. General dorsal color greenish or brownish; belly grayish or brownish.

The young are similar to adults but more conspicuously marked. Length usually 30 to 45 inches, to about 50. Scales keeled; anal plate divided. Scale rows 27 to 29 at midbody.

Variation: No geographic variation is recognized in Kentucky; *Natrix cyclopion cyclopion* (Duméril, Bibron, and Duméril), the common green water snake, is the only subspecies occurring in the state. Females are usually larger than males.

Confusing Species: Resembles several other water snakes, but it alone has some subocular scales interposed between the eye and the upper labials. Sometimes confused with the cottonmouth, which has a pit on each side of the face between the eye and the nostril.

Habitat: Quiet, shallow water—bayous, swamps, sloughs, lakes, impoundments, and sluggish rivers. This is a mud and shore loving species, often seen basking on vegetation along the water's edge.

Kentucky Distribution: Apparently limited to the Jackson Purchase, but the eastern limits of its range are poorly known.

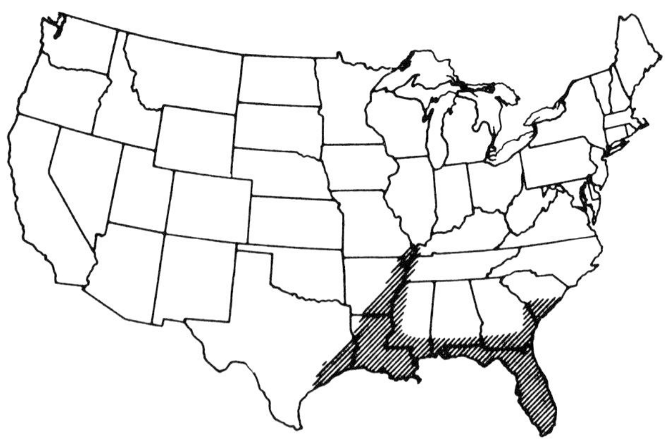

Range of the Green Water Snake

It seems nowhere common, though there are little data on abundance in Kentucky.

Life History: Nothing is known of the life history of these snakes in Kentucky. Farther south, they emerge from hibernation about the middle of March and usually disappear in October. Mating occurs in April, shortly after emergence, and after a gestation period of about 4 months the young are born from late July to early September. The size of the litter is variable, depending generally on the size of the female; an average litter numbers between 10 and 20. The newborn young are quite large, averaging about 9 or 10 inches in length.

Food consists of a variety of vertebrates, with frogs, fishes, toads, and salamanders predominating.

Red-bellied Water Snake LIX

Natrix erythrogaster (Forster)

Recognition: A large, plain, dark gray or greenish to black water snake with an unmarked reddish or yellow belly. Sometimes there are faint black-bordered crossbars on the back, and in some individuals the dark color of the sides extends onto the ends of the belly scutes. Sometimes, too, the lower sides are slightly grayish or even greenish. The young are quite unlike the adults in appearance. They are boldly marked with a series of dark dorsal blotches alternating with smaller lateral blotches on a pinkish or reddish background. The pattern fades with age but may faintly persist into adulthood. Length usually 30 to 48 inches, to 62. Scales keeled; anal plate normally divided (single in about 5 percent of the population). Scale rows 21 to 25 at midbody.

Variation: A highly variable species, varying with both age and geography. Two subspecies are recognized in Kentucky.

Natrix erythrogaster flavigaster Conant. Yellow-bellied Water

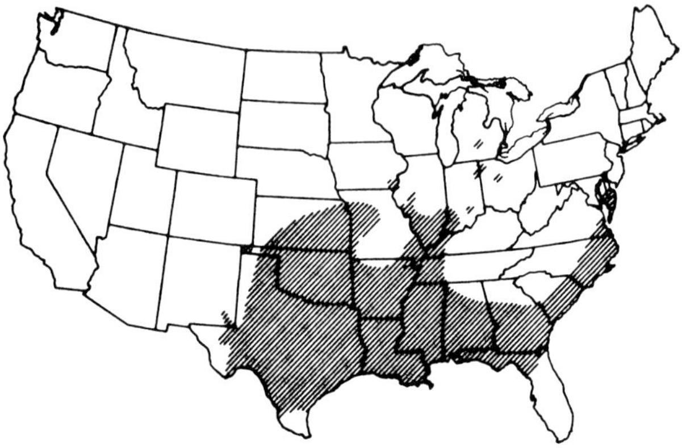

Range of the Red-bellied Water Snake in the United States

Snake. A subspecies of the red-bellied water snake that has a yellow belly. Dorsum and sides grayish or greenish, sometimes with a faint dorsal pattern of dark-edged light transverse bars on the darker background. Belly yellow, often washed with faint orange, with the ends of the scutes occasionally having dark pigment. Occurs in the southern and western sections of the Jackson Purchase.

Natrix erythrogaster neglecta Conant. Northern Copperbelly. Dorsum and sides dark, often black. Belly red, with the dark pigment of the sides often invading the ends of the ventral scutes. Occurs in the central and northeastern sections of the Jackson Purchase and the Land Between the Lakes, thence up the Ohio River at least into Henderson County.

Males have longer tails and more caudals than females, but the overlap precludes reliable separation of the sexes by external features.

Confusing Species: Sometimes other water snakes are uniformly plain above, but they usually have boldly patterned bellies. The young are similar to young diamond-backed water snakes, but the latter have half-moons on the belly, more scale rows, and the dorsal and lateral blotches connected by oblique dark lines.

Habitat: River bottom swamps and other sluggish water. On occasion they congregate about spillways and in summer sometimes wander a considerable distance from the water.

Kentucky Distribution: Throughout the Jackson Purchase and the Land Between the Lakes. Eastern limits of the range are unknown, but it occurs in Hopkins County and ranges up the Ohio River at least into Henderson County.

Life History: Ordinarily this species does not emerge from hibernation before April, but a particularly warm spell in March will cause them to come out temporarily. Mating apparently occurs in May, and the young are born in September or October. They number up to 50 in a litter (20 is about average) and are from about 9 to 12 inches long at birth.

Food of adults consists of fishes, frogs, toads, and salamanders, with catfish and sirens apparently preferred. Occasionally they congregate about a spillway from an impoundment and feed on the fishes washed over the dam.

Kirtland's Water Snake LX

Natrix kirtlandi (Kennicott)

Recognition: A small water snake with 2 rows of alternating squarish blotches on either side of the midline; the spots are sometimes indistinct. Ground color reddish brown; dorsal spots dark brown to black. Belly reddish, with a prominent row of round black spots down either side. The young are dark and virtually unicolored above, and the belly is deep red. Length usually 12 to 18 inches, to 22. Scales keeled; anal plate divided. Scale rows 19 at midbody.

Variation: No geographic variation is recognized. Males have fewer ventrals, more caudals, and longer tails than females. Caudals average about 60 in males and only about 53 in females; the overlap is slight.

Confusing Species: Nearly uniform individuals might be confused with the queen water snake, which has a yellowish belly with 4 stripes. The red-bellied snake has no conspicuous black spots, either above or below.

Habitat: The least aquatic of our water snakes, often found a considerable distance from water. Although able to swim well, it is usually found hiding under stones or debris in wet meadows.

Kentucky Distribution: Known from Jefferson County. Perhaps occurs in many of the Ohio River Valley counties from Daviess to Campbell. It now seems nowhere common.

Life History: Little is known of the life history of these snakes in Kentucky. They apparently emerge from hibernation in late March or April and mate in early May. The young,

Range of Kirtland's Water Snake

numbering up to 22, are born in August and September and range from 5 to 6¾ inches in length. Food consists of earthworms, insects, slugs, frogs, toads, salamanders, and minnows.

Remarks: When first encountered, this small snake often flattens its body and remains still until touched; then it reacts by writhing violently. Any accurate observation on the life history or distribution of this species would be valuable.

Diamond-backed Water Snake LX

Natrix rhombifera (Hallowell)

Recognition: A large, heavy-bodied water snake with an obvious pattern on the back and sides consisting of dark rectangular black or dark brown dorsal blotches connected by oblique dark lines from their corners with the alternating dark, broad, lateral bars. The ground color is a lighter brown or dirty yellowish. The general impression is of a dark chainlike pattern on a lighter background. The light spaces between the darker dorsal blotches are frequently diamond-shaped. The belly is yellowish, marked with dark half-moons or spots, especially near the ends of the scutes. Youngsters are more brightly colored and have slenderer heads; frequently the belly is reddish. Length usually 30 to 48 inches, to 67. Scales keeled; anal plate divided. Scale rows 25 to 31 at midbody.

Variation: No geographic variation is recognized in Kentucky; *Natrix rhombifera rhombifera* (Hallowell), the common diamond-backed water snake, is the only subspecies occurring in the state. Males are smaller than females with slightly longer tails. Older males are unique among our snakes in possessing a series of papillae beneath the chin.

Confusing Species: Most like the other patterned water snakes, but the pattern exhibited by this species is distinctive.

It might be confused with the cottonmouth, but the latter possesses a sensory pit on either side of the head.

Habitat: Aquatic, widely varied, from small streams and waterholes to large rivers and lakes. It may be observed basking on vegetation or piles of brush along sluggish watercourses or in swamps.

Kentucky Distribution: Throughout the Jackson Purchase. The eastern limits are unknown, but it ranges up the Ohio River probably at least to Daviess County. It likely occurs in the lower reaches of the Green, Pond, and Tradewater rivers in western Kentucky as well as in assorted swamps and sloughs in that region. It apparently reaches its greatest abundance along the permanent streams, sloughs, and swamps of the Jackson Purchase.

Life History: Little is known of the natural history of these snakes in Kentucky. They are usually out of hibernation and active by mid-April and apparently mate soon after emergence.

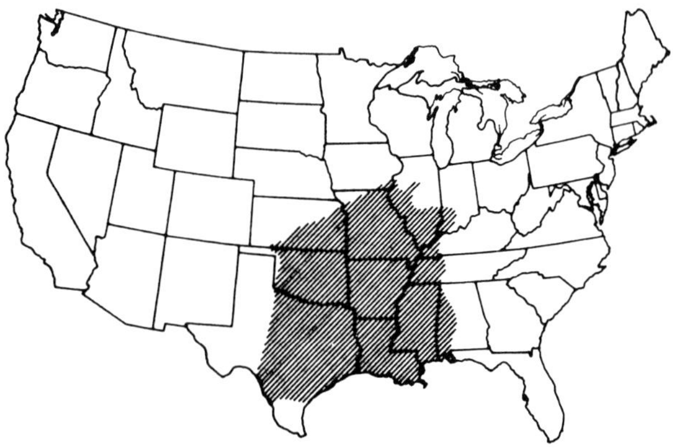

Range of the Diamond-backed Water Snake in the United States

The young, numbering from 18 to 62, are usually born from late August to October and are about 10½ inches long at birth.

This species is most active by night, usually spending the day in some secluded spot. Evil-tempered, it will bite viciously when captured and usually void quantities of a vile-smelling thick fluid from the cloaca.

Food consists mainly of fishes, frogs, and toads. Tadpoles, salamanders, crayfish, and young turtles are also frequently eaten.

Common Water Snake LXI

Natrix sipedon (Linnaeus)

Recognition: A large, grayish to dark water snake with dark crossbands on the neck and front of the body and alternating dorsal and lateral blotches on hind part of body; there is a total of more than 20 bands and dorsal spots. Old individuals may appear uniformly colored, but placing them in water will usually enable one to discern the markings. Ground color varies from light gray to dark brown; the markings from bright reddish brown to black. Often with black or reddish half-moons on the belly. The young are usually much more boldly patterned than adults. The markings are almost black, and the ground color usually light gray or almost white. Occasionally an old adult loses the pattern completely and becomes uniformly dark. Length usually 24 to 48 inches, to 58. Scales keeled; anal plate divided. Scale rows about 23 at midbody.

Variation: Highly variable, with age and with geography. Two subspecies are recognized in Kentucky.

Natrix sipedon sipedon (Linnaeus). Northern Water Snake. Dark markings on the back and sides wider than the spaces between them. Belly markings often half-moon-shaped but frequently broken up into dusky mottling. Ranges throughout the Eastern Coal Field and westward to about a line connecting Jefferson and Todd counties where it meets and intergrades across a wide band with:

Natrix sipedon pleuralis Cope. Midland Water Snake. Dark markings on the sides and back narrower than the spaces between them. Belly markings usually in the form of paired half-moon-shaped blotches. Occurs in the Jackson Purchase and the Land Between the Lakes and eastward where it meets and intergrades with the preceding.

Males have relatively longer tails than females and more caudals. The caudals average about 75 in males and 63 in females and usually serve to separate the sexes.

Confusing Species: Similar to the diamond-backed water snake, which has oblique bars connecting the dark dorsal and lateral blotches. Like the broad-banded water snake, which has a dark line from the eye to the angle of the jaw, from 11 to 17 crossbands, and usually no alternating lateral and dorsal blotches. Often confused with the venomous cottonmouth, but the latter has facial pits which all water snakes lack.

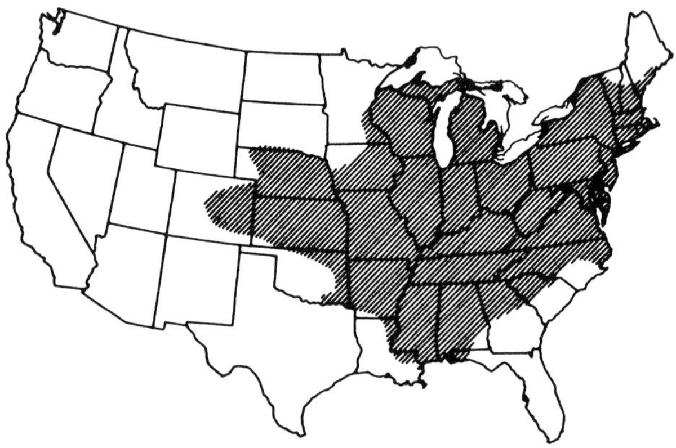

Range of the Common Water Snake in the United States

Habitat: Nearly any water, from springs and farm ponds to the largest lakes, and from small woodland streams to the largest rivers; however, it seems partial to flowing water. It is fond of basking and is often seen lying on piles of drift or on overhanging trees and bushes.

Kentucky Distribution: Statewide and abundant. This is the only species of blotched water snake across most of the state.

Life History: These snakes usually emerge from hibernation in late March or April and mate soon thereafter. The young, numbering up to 70 in a litter, are usually born in late August, September, or early October; at birth they are usually between 7½ and 9 inches long.

A female selects a sheltered spot, such as under a large flat stone, to give birth to her youngsters. There is no parental care, and the mother and young soon go their separate ways.

Adults normally spend the daylight hours hiding or basking on some elevated spot and forage by night. They are difficult to approach by day but can be captured rather easily at night.

Food consists of fish, frogs, tadpoles, salamanders, and the like.

Banded Water Snake LXI

Natrix fasciata (Linnaeus)

Recognition: A large, heavy-bodied water snake with a pattern of from 11 to 17 broad, dark crossbands on the body. There is a dark stripe from the eye to the angle of the jaw and the belly has dark, squarish markings. Color variable; crossbands usually from reddish-brown to black on a yellowish ground color. The young are more boldly marked, with the ground color a clearer yellow and the crossbands darker than adults. Length usually 20 to 36 inches, to about 45. Scales keeled; anal plate divided. Scale rows 21 to 25 at midbody.

Variation: *Natrix fasciata confluens* Blanchard, the broad-banded water snake, is the only subspecies occurring in the state. There is considerable individual variation in color and pattern. Males have longer tails and a higher number of caudal scutes than females. The caudals average about 72 in males, 62 in females; usually the sexes can be separated by this characteristic.

Confusing Species: Like our other banded water snakes, but unique in having so few and such broad crossbands on the body. It alone of our water snakes has a dark line from the eye to the angle of the jaw.

Habitat: Large shallow water areas. Although sometimes inhabiting sluggish streams, it more commonly occurs in cypress swamps, marshes, and lakes.

Kentucky Distribution: Known in Kentucky only from the swamps and sloughs along the Mississippi River in the western portion of the Jackson Purchase. It seems nowhere abundant.

Life History: Little data are available on this species in Kentucky. They usually emerge from hibernation in late March

Range of the Banded Water Snake

and April and apparently mate shortly thereafter. Young, numbering up to about 50, are born in August and September and are about 7 to 9½ inches at birth.

Food consists of vertebrates; frogs seem preferred, but fishes and salamanders are taken at times.

Remarks: So little is known of this species in Kentucky that any accurate observations on its distribution, habitat, or habits would be of value.

FAMILY VIPERIDAE Vipers

These are venomous snakes with the whole venom apparatus developed to its maximum. The maxillary bones are short and deep and lie at the front of the mouth. Attached to their posterior margins is a pair of large, highly specialized teeth, sometimes grooved, sometimes hollow, for the injection of venom. When at rest the fangs are folded back along the roof of the mouth, but when the mouth is opened to strike, the maxillae rotate forward, erecting the fangs.

The venom-secreting glands are modified salivary glands lying just below and behind the eye. Each is connected by a duct to the hollow of the fang, and when the fang is inserted a muscular contraction injects the venom. The venom itself is a complex mixture of proteins and enzymes. Some venoms act principally in the destruction of red blood cells and capillary walls; others act mainly on the nervous system. Generally, venom of the vipers is more hemolytic than neurotoxic.

There are two subfamilies of Viperidae: the Viperinae, or true vipers, and Crotalinae, the pit vipers. Ten genera of true vipers are known, confined to Africa and Eurasia, and some of them are among the most deadly of snakes.

The Crotalinae have the maxillary bone hollowed out above by a pit opening between the eye and the nostril. The pit is heat sensitive and is used to locate warm-blooded prey on which the animals feed. Pit vipers occur from the Indo-Australian Archipelago northward to Japan and westward across Asia to Eastern Europe. They are most abundant, however, in the Western Hemisphere, widely distributed across North, Central, and South America.

Seventeen species of Crotalinae, in 3 genera, occur in the United States. All 3 genera, but only 4 species extend into Kentucky.

KEY TO THE KENTUCKY VIPERIDAE

1 a. With a rattle on the end of the tail — 2

b. Tail pointed at end; no rattle — 4

2 a. With 9 large scales (plates) on the crown of the head; tail skinny and rattle minute / *Sistrurus miliarius streckeri* Gloyd, Western Pigmy Rattlesnake (p. 312)

b. With a mixture of large plates and small scales on the crown of the head; tail not skinny; rattle not minute (*Crotalus horridus*) — 3

3 a. With a distinct dark postocular stripe, contrasting with light background; a brown or reddish-brown mid-dorsal stripe on anterior part of body; scale rows 25 / *Crotalus horridus atricaudatus* Latreille, Canebrake Rattlesnake (p. 314)

b. Without stripes as above; head usually unmarked; scale rows 23 / *Crotalus horridus horridus* Linnaeus, Timber Rattlesnake (p. 314)

4 a. Dorsal pattern of 10 to 15 wide dark crossbands on an olive or brown ground color; old individuals have an obscure pattern and are very dark, often black / *Agkistrodon piscivorous leucostoma* (Troost), Western Cottonmouth (p. 304)

2a

2b

3a

4a

4b 

b. Dorsal pattern of 15 to 25 chestnut brown crossbands on a lighter ground color, each constricted toward the midline of the body (*Agkistrodon contortrix*, Copperhead) 5

5 a. Narrow portion of crossbands rarely more than 2 to 3 scales wide, or sometimes failing to meet in the midline; coloration pale / *Agkistrodon contortrix contortrix* (Linnaeus), Southern Copperhead (p. 309)

b. Narrow portion of crossbands 3 to 5 scales wide, all usually meeting at the midline; coloration darker / *Agkistrodon contortrix mokeson* (Daudin), Northern Copperhead (p. 309)

Cottonmouth LXII

Agkistrodon piscivorous Lacépède

Recognition: A large, dark, heavy-bodied, flat-headed, venomous, aquatic snake. Although sometimes so dark as to appear almost black, it usually has a series of broad, dark-edged crossbands across the back that extend down the sides to the belly. The ground color is reddish brown in the young, dusky brown in subadults, and dark olive in adults. The snout is normally light, and there is usually a broad, dark line through the eye. Above and especially below this dark line the side of the face is usually light. The belly is light, heavily blotched with dark brown or black.

The young at birth are brightly colored, the top of the head especially so, strongly suggesting a copperhead. The last inch or so of the tail is bright yellow. This yellow generally is lost dorsally by the time the animal reaches a length of 16½ inches but remains ventrally, though less brilliant, until the animal attains a length of over 28 inches. The brillance of color exhibited at birth is maintained with little change throughout the first year of life, but afterwards it gradually darkens until the older animals are quite dark. In most individuals, however, indications of the dorsal pattern persist and can be discerned, especially when the animal is wet.

Length usually 20 to 36 inches, to 54. Scales weakly keeled; anal plate single. Scale rows usually 25 (23 to 26) at midbody.

Variation: No geographic variation is recognized in Kentucky; *Agkistrodon piscivorous leucostoma* (Troost), the western cottonmouth, is the only subspecies occurring in the state. Males on the average have more caudal scutes and longer tails than females, but the overlap is so great that the sexes cannot reliably be separated by external features.

Confusing Species: Similar in general appearance to several species of large water snakes of the genus *Natrix*. Even at a distance, however, the light upper jaw of the cottonmouth is distinctive. An undisturbed cottonmouth, swimming leisurely, usually has the anterior 2 or 3 inches of the body angled upward out of the water, with the head bent slightly downward, paralleling the surface. Snakes of the genus *Natrix* swim with the head flat on the surface. These two characteristics are ordinarily sufficient to enable an experienced observer to identify a cottonmouth from a distance of several yards. At one time or another, surely every snake found near the water has been confused with the cottonmouth. By far the best means to separate the cottonmouth from other snakes is to learn its features and pattern. The facial pit between the eye and nostril, the vertical or elliptical pupil, and the essentially single row of scales under the tail immediately identify the cottonmouth as a pit viper. The 12 to 18 broad, usually dim crossbands and the broad, dark line through the eye separate it readily from the copperhead; our only other pit vipers have rattles.

Habitat: Cypress swamps, oxbow lakes, bayous, and sluggish streams. In autumn it leaves the swamps and goes up into wooded bluffs to hibernate in holes in the ground.

Kentucky Distribution: Present valid records show the cottonmouth to be confined to the western end of the state, reaching its greatest abundance in the Jackson Purchase. It surely occurs at least sparingly in every county in the Purchase and extends eastward as far as Muhlenberg County. There are numerous areas along the Green River and its tributaries below Morgantown and along the Tradewater River that are apparently good habitats. Cottonmouths should be expected in almost any extensive swampland in the state west of a line connecting Warren and Grayson counties, but not to the east.

Because of its restricted habitat and limited distribution, the cottonmouth should not be considered a common Kentucky snake. In places, however, the species reaches a population density exceeded by no other snake in the state. Counts in several small areas of approximately 1,000 square feet each indicate a population in selected areas exceeding 300 snakes to the acre. Once 5 of us caught 50 cottonmouths in 2 hours. No count was kept of those we saw and did not catch. On occasion as many as 4 or 5 snakes can be seen sunning themselves on the same pile of drift. Such spots are certainly limited and difficult to reach; seldom are they visited.

Life History: In late August and early September the snakes begin a rather leisurely exodus from the swamplands to adjacent low wooded hillsides. Here they go into any suitable hole in the ground to hibernate.

With the advent of warm weather, they return to the swamps. A few consecutive warm days in late March will set off the migration, but they sometimes do not emerge before April. In mid-April a few youngsters born the preceding fall may still be found in the woods under logs and stumps or even actively on their way to the water, but the larger snakes are generally well distributed in the swamps by that time. The majority of the youngsters are confined to the first 100 yards of swampland immediately bordering the hillsides. By late April the youngsters are farther out in the swamps and by June are widely distributed.

Mating apparently occurs shortly after emergence in the spring, and the 2 to 11 young are usually born in early September. At hatching they are from 7 to 10½ inches long. There is some evidence that the females leave the swamps to bear their young on the hillsides where they hibernate. Although

Range of the Cottonmouth

the situation is not clear, it seems that the young do not enter the swamps until the following spring.

During the day the cottonmouth is mostly inactive, lying sometimes coiled, sometimes not, in shaded areas when the sun is shining brightly, and in the sunlight when the days are cool. They are frequently to be found on the mud at the edge of a pool or slough but more often on some pile of emergent drift or vegetation. The smaller snakes often climb on cypress knees where they are surprisingly difficult to see. During the day the snakes are not easily disturbed and will generally lie perfectly still and allow one to approach within 2 or 3 feet. They do not often open their mouths to display the "cottonmouth," but an occasional individual will do so. The larger individuals seem more easily alarmed than the smaller ones.

Examination of 42 cottonmouths from Kentucky that contained identifiable food remains revealed that *Siren intermedia* was the most abundant food item by both volume and occurrence. Frogs of the genus *Rana* ranked second. Together these two items comprised almost two-thirds of the food of the snakes. The other food items were distributed among the fishes, reptiles, and other amphibians.

In captivity they take mice readily, but no mammals have been found in specimens from Kentucky, likely because small mammals are rare in the swampy areas normally inhabited by the snakes.

The cottonmouth, in contrast to the copperhead and rattler, frequently retains a grip on animals struck. This is perhaps an adaptation to a semiaquatic existence, where it would be difficult or impossible to trail wounded prey.

Active foraging for food begins in the late afternoon or early evening and continues well into the night. Possibly those snakes whose foraging is not successful continue throughout the night and into the next morning, for empty snakes have been seen foraging shortly after daybreak.

While swimming in search of food, the cottonmouth holds its head and the anterior 2 or 3 inches of the body out of the water as if it were looking about. When disturbed, it lowers the head and body to the water and swims rapidly away, sometimes on the surface, sometimes diving. They will take food under the water, on the surface, on land, and will even climb onto vegetation for it.

Remarks: If the snakes were a little more aggressive, they might inflict some damage to man himself and particularly his livestock, but the cottonmouth is quite docile. One farmer I know in the Jackson Purchase maintains a herd of cattle that roams at will in areas where cottonmouths are abundant and has never had one of his cattle bitten by a snake.

Copperhead LXIII

Agkistrodon contortrix (Linnaeus)

Recognition: A medium-sized, venomous snake with a coppery-red head and a dorsal series of hour-glass-shaped, chestnut-colored blotches. In some individuals the markings are quite narrow in the mid-dorsal region, and in a few the marks on the sides fail to connect across the back. Ground color light reddish-brown through yellowish to dark brown. Belly light yellowish-brown or reddish-brown, with brown blotches on the ends of the ventrals alternating with dorsal crossbands. The young are paler than adults and have a narrow dark streak through the eye that separates the light cheek from the dark head. The tail tip of newborn young is bright yellow. Length usually 20 to 36 inches, to 53. Scales weakly keeled; anal plate single. Scale rows usually 23 (23 to 25) at midbody.

Variation: Two subspecies of copperheads are recognized in Kentucky.

Agkistrodon contortrix contortrix (Linnaeus). Southern Copperhead. Dorsal saddles frequently broken in the mid-dorsal region and meeting quite narrowly (2 to 3 scales wide) when they do connect. This results in a somewhat more "wasp-waisted" hourglass than that found on the northern copperhead. Ranges about the lowlands and cypress swamps of the southern and western areas of the Jackson Purchase. Meets and intergrades in a broad band across the eastern portion of the Jackson Purchase and the western end of the Mississippian Plateau with:

Agkistrodon contortrix mokeson (Daudin). Northern Copperhead. This is the common copperhead of Kentucky. The "hour-glasses" usually are complete, the two lateral halves meeting broadly (3 to 5 scales wide) in the mid-dorsal region. Darker, more richly colored, less pale and pink than the southern copperhead. It ranges from the area of intergradation eastward in suitable habitats across the remainder of the state.

Males usually average more caudals and a longer tail than females, but the overlap is so great as to preclude the reliable separation of the sexes by external features.

Confusing Species: Most like the milk snake, in which the dark crossbands are widest in the mid-dorsal region and the

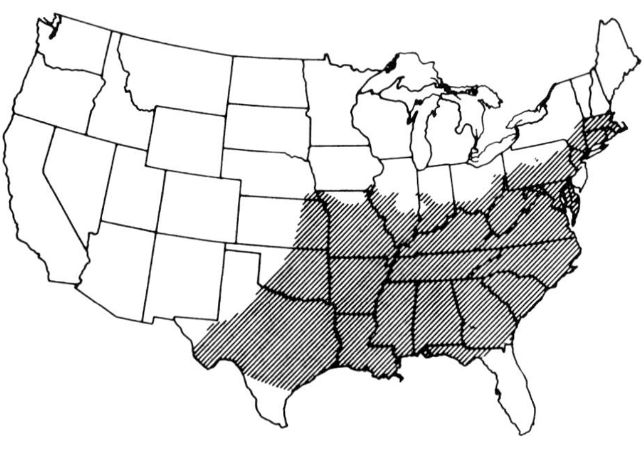

Range of the Copperhead in the United States

belly has black, squarish blotches. Water snakes have strongly keeled scales. Corn snakes have checkerboard bellies. At one time or another nearly every snake has been confusd with the copperhead. The surest identification procedure is to learn the features and pattern of the copperhead: facial pits, copper-colored head, hour-glass-shaped dark crossbands on the body, scutes on the underside of the tail in a single row, and vertical pupils.

Habitat: Woodlands, especially wooded hillsides or bluffs with rock outcrops. It is frequently encountered about old houses and barns, probably because of the abundant food and cover. In summer one is sometimes found in an open field or along a stream.

Kentucky Distribution: Statewide in suitable habitats. The copperhead has been extirpated in most urban areas, and it seems absent in most of the Inner Blue Grass. It occurs sparingly along the wooded bluffs along the course of the Kentucky River through the area, however.

Copperheads are more numerous than would appear to the casual observer. They are almost always quite docile and either lie quietly or slither away when approached. This habit, coupled with their excellent concealing coloration, renders them quite difficult to see. As an example of their abundance in some places, 7 adults were captured in less than 15 minutes one evening from an area less than 25 feet square in Breathitt County.

Life History: Copperheads are sometimes active by early April, but are more commonly encountered later in the season. They remain active into October or even November in warm seasons. They mate soon after emergence, in April or May, and the females give birth to from 2 to 11 young, usually in August or early September, but sometimes as early as July. The newborn youngsters are 8 or 9 inches long.

Food of the copperhead in Kentucky consists of a variety of animals, but small mammals are the major items of diet. Caterpillars are frequently eaten, as are lizards, and there is a record of one copperhead from Breathitt County whose stomach was stuffed with 17-year cicadas.

Pigmy Rattlesnake LXIV

Sistrurus miliarius (Linnaeus)

Recognition: A small, grayish rattlesnake with a tiny rattle and a series of 9 large plates on the top of the head. Ground color usually grayish-brown. A series of irregular dark mid-dorsal spots often form transverse bars. There are 1 or 2 rows of dark spots on each side of the body and usually a reddish mid-dorsal stripe. The belly is light brownish or cream, marked with darker brown or black blotches. The young are considerably lighter than adults. Length usually 15 to 20 inches, to about 25. Scales strongly keeled; anal plate single. Scale rows 21 at midbody.

Variation: No geographic variation is recognized in Kentucky nor is any likely to occur because of the limited distribution of this species in the state. *Sistrurus miliarius streckeri* Gloyd, the western pigmy rattlesnake, is the only subspecies known in Kentucky. Males have fewer ventrals, more caudals, and a relatively longer tail than females, but the overlap is such as to preclude reliable separation of the sexes by external features.

Confusing Species: Our only other rattlesnake, *Crotalus horridus*, has a combination of a few large and many small scales on the crown of the head, and a much larger, more massive rattle.

Habitat: Varied, but apparently limited to areas near water.

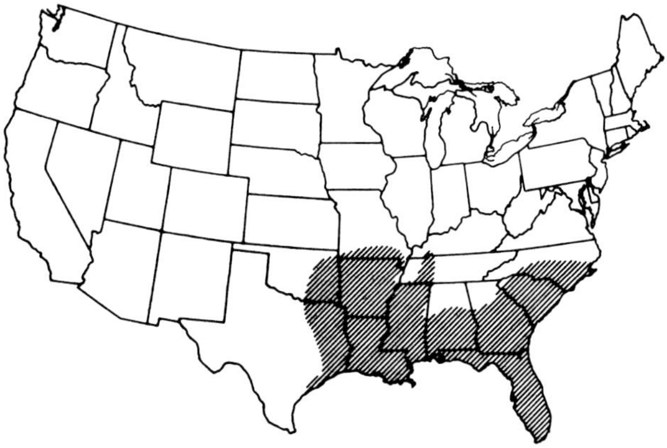

Range of the Pigmy Rattlesnake

It occurs in floodplains, wet flatwoods, swamps, and marshes as well as in the drier wooded uplands.

Kentucky Distribution: Currently this snake is known to occur in Kentucky only along the southeastern edge of the Jackson Purchase in the Land Between the Lakes. That it occurs on the west shore of Kentucky Lake in Calloway County is suggested by an individual found on the west shore in Tennessee, about 2 miles south of the Kentucky-Tennessee border.

Life History: There are essentially no data on the life history of this tiny rattlesnake in Kentucky. The young are born alive, up to 18 or so per litter, usually in late July or early August. They are some 5 to 6 inches long at birth and capable of inflicting a nasty bite.

Mice constitute a large part of the diet, but birds, frogs, lizards, and sometimes other snakes are eaten.

Remarks: The little pigmy rattlesnake makes up in aggressiveness what it lacks in size; it is a fiery-tempered little beast.

Although there seem to be no records of fatalities from its bite, it should be treated with considerable respect, for it is capable of inflicting a serious wound.

Rattlesnake LXIV

Crotalus horridus Linnaeus

Recognition: A large, heavy-bodied, banded, venomous snake with a rattle on the tail. It occurs in 2 color phases: yellow and black. In the yellow phase the ground color is yellow, brown, or grayish with darker crossbands. In the black phase the ground color is masked by a dark pigment, and the still darker crossbands are not so conspicuous. On occasion a beautifully marked, velvety, olive-green animal is encountered; totally black specimens are not uncommon in some areas. Despite popular opinion these color differences have no relation to the sex of the individual.

Some rattlesnakes have a broad, dark stripe running backward from the eye through the angle of the jaw, while others lack the stripe. Occasionally there is a reddish mid-dorsal stripe. The belly is yellowish, thickly sprinkled with black. The young are more contrastingly colored than adults, having a lighter ground color. They do not possess a rattle but merely a "prebutton" or "button," depending upon their age. Length usually 30 to 60 inches, to 74½ inches (there are questionable records of as long as 108 inches). Scales strongly keeled; anal plate single. Scale rows 23 to 25 at midbody.

Variation: In addition to the individual variation that is not correlated with age, sex, or geography, there is considerable geographic variation in the rattlesnakes across Kentucky. Two subspecies are recognized in the state.

Crotalus horridus horridus Linnaeus. Timber Rattlesnake. This race usually has neither a light reddish mid-dorsal stripe nor a broad, dark stripe running backward from the eye. Scale rows usually 23. Ranges across eastern and central Kentucky to about the eastern edge of the Jackson Purchase where it meets and intergrades broadly with:

Crotalus horridus atricaudatus Latreille. Canebrake Rattlesnake. With a broad, dark stripe running backward from the eye across the angle of the jaw and beyond. Usually, there is a reddish mid-dorsal stripe, at least anteriorly, that is about 3 scales wide. Scale rows usually 25. Ranges in the southern and western sections of the Jackson Purchase eastward to meet and intergrade with the preceding at about the western edge of the Mississippian Plateau.

Males usually have fewer ventrals, more caudals, and relatively longer tails than females, but the overlap is such as to preclude the reliable separation of the sexes by external features.

Habitat: The timber rattlesnake is an inhabitant of the woodlands, most commonly encountered on dry ridges with rock outcrops. The canebrake rattlesnake is an inhabitant of moist woodlands, inhabiting the floodplains, canebrakes, and swamplands of the south. Rattlesnakes seem nowhere common in Kentucky. They have been extirpated from urban areas and

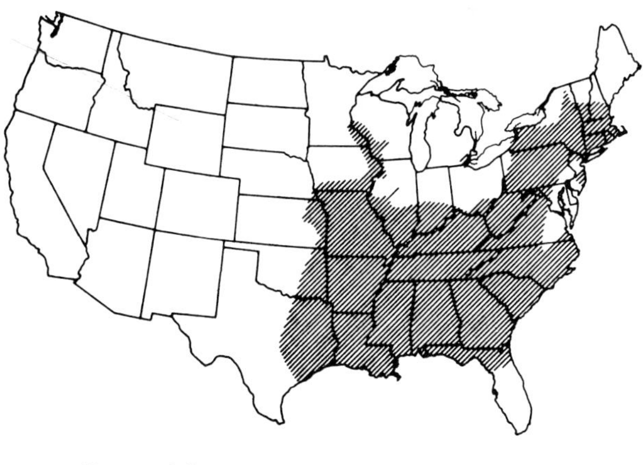

Range of *Crotalus horridus* in the United States

are scarce or absent throughout the Blue Grass and the more intensively farmed sections of the Mississippian Plateau. They are probably most common in the southeastern half of the Eastern Coal Field.

Life History: Rattlesnakes sometimes congregate in considerable numbers about favored den sites in fall prior to going into hibernation. Such sites are usually wooded bluffs or hillsides with rock outcrops containing numerous deep crevices extending well below frost line. Some sites are used year after year and have a known history as a "snake den" going back many years.

Rattlesnakes usually emerge in April or May and mate soon after emergence, sometimes before leaving the den. The females give birth to up to 17 young (usually 6 to 10) in late August or September. The youngsters are variable in size, from about 7 to 12 inches long.

Their diet consists almost wholly of small mammals, mostly mice of various species; they will, however, sometimes take a bird. Chipmunks are frequently eaten, and on two separate occasions an adult gray squirrel has been taken from the stomach of a large Kentucky rattlesnake.

Remarks: The rattle, the most characteristic feature of a rattlesnake, is one of the strangest structures in nature; nothing remotely resembling it is found anywhere else in the animal kingdom. It is a series of interlocking, dry, horny segments, flattened from side to side, that creates a buzzing noise when shaken vigorously. It is shaken, incidentally, from side to side. At birth, a young rattlesnake possesses a "prebutton." Within a week or so it sheds its skin and "prebutton" and acquires a "button." Henceforth, throughout its life, it acquires a new segment every time the skin is shed (2 to 4 times per year). Terminal segments are frequently lost, and it is rare that a snake is found with more than 12 or 15 segments. Obviously, there is no correlation between the age of the snake and the number of segments of the rattle.

Much speculation exists as to the origin and function of the rattle without any satisfactory answer. However, all our pit vipers, as well as many nonvenomous snakes, will rapidly vibrate the tail tip when disturbed, exactly like a rattlesnake. The rattler vibrates its tail for the same reason other snakes do; it is nervous, or angry, or frightened. The fact that a buzzing noise results is completely incidental and quite unknown to the snakes; they are as deaf as posts.

LIST OF PLATES (REPTILES)

(Above) Common Snapping Turtle, *Chelydra serpentina serpentina*

(Below) Alligator Snapping Turtle, *Macroclemys temmincki*

Head of Alligator Snapper showing "worm"

(Above) Stinkpot, *Sternotherus odoratus*

(Below) Eastern Mud Turtle, *Kinosternon subrubrum subrubrum*

(Above) Eastern Box Turtle, *Terrapene carolina carolina*

(Below) Map Turtle, *Graptemys geographica*

(Above) Ouachita False Map Turtle, *Graptemys pseudogeographica ouachitensis*

(Below) Mississippi Map Turtle, *Graptemys kohni*

(Above) Midland Painted Turtle, *Chrysemys picta marginata*

(Below) Red-eared Turtle, *Chrysemys scripta elegans*

XXXVIII

(Above) Hieroglyphic Turtle, *Chrysemys concinna hieroglyphica*

(Below) Missouri Slider, *Chrysemys floridana hoyi*

(Above) Common Smooth Softshell Turtle, *Trionyx muticus muticus*

(Below) Eastern Spiny Softshell Turtle, *Trionyx spinifer spinifer*

XL

(Above) Northern Fence Lizard, *Sceloporus undulatus hyacinthinus*

(Below) Ground Skink, *Lygosoma laterale*

(Above) Five-lined Skink, *Eumeces fasciatus*

(Below) Broad-headed Skink, *Eumeces laticeps*

XLII

(Above) Southeastern Five-lined Skink, *Eumeces inexpectatus*

(Below) Northern Coal Skink, *Eumeces anthracinus anthracinus*

(Above) Six-lined Racerunner, *Cnemidophorus sexlineatus sexlineatus*

(Below) Eastern Slender Glass Lizard, *Ophisaurus attenuatus longicaudus*

XLIV

(Above) Eastern Worm Snake, *Carphophis amoenus amoenus*

(Below) Western Mud Snake, *Farancia abacura reinwardti*

Northern Ringneck Snake, *Diadophis punctatus edwardsi*

(Above) Eastern Hognose Snake, *Heterodon platyrhinos platyrhinos*

(Below) Eastern Rough Green Snake, *Opheodrys aestivus aestivus*

(Above) Northern Black Racer, *Coluber constrictor constrictor* (adult)

(Below) Northern Black Racer, *Coluber constrictor constrictor* (young)

Eastern Coachwhip, *Masticophis flagellum flagellum*

(Above) Corn Snake, *Elaphe guttata guttata*

(Below) Corn Snake, *Elaphe guttata guttata* (head)

L

(Above) Black Rat Snake, *Elaphe obsoleta obsoleta* (adult)

(Below) Black Rat Snake, *Elaphe obsoleta obsoleta* (young)

(Above) Northern Pine Snake, *Pituophis melanoleucus melanoleucus*

(Below) Prairie Kingsnake, *Lampropeltis calligaster calligaster*

LII

Black Kingsnake, *Lampropeltis getulus niger*

→

(Above) Eastern Milk Snake, *Lampropeltis doliata triangulum* (adult)

(Middle) Eastern Milk Snake, *Lampropeltis doliata triangulum* (young)

(Below) Scarlet Kingsnake, *Lampropeltis doliata doliata*

(Above) Northern Scarlet Snake, *Cemophora coccinea copei*

(Below) Southeastern Crowned Snake, *Tantilla coronata coronata*

(Above) Eastern Ribbon Snake, *Thamnophis sauritus sauritus*

(Below) Western Ribbon Snake, *Thamnophis proximus proximus*

(Above) Eastern Garter Snake, *Thamnophis sirtalis sirtalis*

(Below) Eastern Earth Snake, *Virginia valeriae valeriae*

(Above) Midland Brown Snake, *Storeria dekayi wrightorum*

(Below) Northern Red-bellied Snake, *Storeria occipitomaculata occipitomaculata*

LVIII

(Above) Queen Water Snake, *Regina septemvittata septemvittata*

(Below) Common Green Water Snake, *Natrix cyclopion cyclopion*

(Above) Yellow-bellied Water Snake, *Natrix erythrogaster flavigaster* (adult)

(Below) Yellow-bellied Water Snake, *Natrix erythrogaster flavigaster* (young)

Kirtland's Water Snake, *Natrix kirtlandi*

Common Diamond-backed Water Snake, *Natrix rhombifera rhombifera*

(Above) Northern Water Snake, *Natrix sipedon sipedon*

(Below) Broad-banded Water Snake, *Natrix fasciata confluens*

LXII

(Above) Western Cottonmouth, *Agkistrodon piscivorous leucostoma* (adult)

(Below) Western Cottonmouth, *Agkistrodon piscivorous leucostoma* (young)

Northern Copperhead, *Agkistrodon contortrix mokeson*

LXIV

(Above) Western Pigmy Rattlesnake, *Sistrurus miliarius streckeri*

(Below) Timber Rattlesnake, *Crotalus horridus horridus*

ADDITIONAL SOURCES OF INFORMATION

For the convenience of those who may wish to read further, I have listed below five basic sources of information on Kentucky herpetology.

1. Academies of Science. These are of various sizes and states of affluence, but most publish a journal, variously titled and usually quarterly. The state academies of Illinois, Indiana, Kansas, Kentucky, Michigan, Ohio, Tennessee, and West Virginia have all published papers dealing in whole or in part with Kentucky herpetofauna. The Chicago Academy and the Biological Society of Washington have also published such papers.

2. Journals. These are the publications, usually issued quarterly, of the various professional societies. Some deal exclusively with herpetology, some include fishes, others are more inclusive, and one, at least, deals with any aspect of science. They constitute one of the best sources of reliable, up-to-date information. The back issues, on file in many libraries, constitute a veritable mine of information. *Copeia, Ecological Monographs, Ecology, Herpetologica, Journal of Herpetology,* and *Science* are some of the more important ones frequently containing data on Kentucky herpetology. Although not a journal in this sense, the *American Midland Naturalist* is an excellent source of information.

3. Museums. Many state and local museums have published information on Kentucky species. Among the more important ones are the American Museum, the Field Museum, the Florida State Museum, the New York State Museum, and the United States National Museum.

4. State Publications. Publications of the various state conservation departments often contain pertinent information. Those of the Illinois Natural History Survey are outstanding. In past years the Kentucky Geological Survey has published data on Kentucky vertebrates.

5. Universities. I have chosen to group university publications into three broad categories: university press publications, theses, and miscellaneous publications.

A. University press publications. The Comstock Publishing Company and its successor, Comstock Publishing Associates, a division of the Cornell University Press, is preeminent in this field by virtue of the "Handbook" series edited by A. H. Wright. Salamanders, frogs and toads, turtles, lizards, and snakes are all treated in separate volumes. The presses of the universities of Missouri and Tennessee have each published books that have data on Kentucky species.

B. Theses. Theses dealing with some phase of herpetology are sequestered away in the libraries of nearly every university offering graduate degrees in biology. Much of the information in them is published, but in all too many cases much valid data remains buried. As a case in point, much distribution data on Kentucky species lies unpublished in a thesis at George Peabody College for Teachers. The University of Kentucky, the University of Louisville, and the University of Cincinnati all have essentially unpublished theses containing data on Kentucky herpetofauna.

C. Miscellaneous university publications. There are a number of series of biological publications issued by various universities. An outstanding publication, the *American Midland Naturalist*, containing many articles dealing in whole or in part with Kentucky herpetofauna, is published by the University of Notre Dame. The publications of the Museum of Natural History of the University of Kansas and the *Occasional Papers* of the Museum of Zoology of the University of Michigan have been outstanding for many years and contain a great deal of pertinent information. The *Tulane Studies in Zoology* is another important source.

SELECTED BIBLIOGRAPHY

Most of the publications in this list deal at least in part with distribution and/or life history of amphibians and reptiles in Kentucky. Others are included if they contain substantial or particularly pertinent information on species occurring in Kentucky. No attempt has been made for completeness, although by far the majority of the pertinent publications on Kentucky herpetology are included.

Adams, Wm., Jr. 1939. Some amphibia of eastern Kentucky. M.S. Thesis. George Peabody College for Teachers. 125 pp.

Anderson, Paul. 1966. The reptiles of Missouri. University of Missouri Press, Columbia. 330 pp.

Bailey, Vernon. 1933. Cave life of Kentucky. Amer. Midland Nat. 14 (5): 383-635.

Barbour, Roger W. 1941. *Gastrophryne carolinensis* in Kentucky. Copeia, No. 4, p. 262.

———. 1946. Additional record of the narrow-mouthed toad in Kentucky. Copeia, No. 1, p. 44.

———. 1950. The distribution of *Carphophis amoena* in Kentucky. Copeia, No. 3, p. 237.

———. 1950. A new subspecies of the salamander *Desmognathus fuscus*. Copeia, No. 4, pp. 277-78.

———. 1950. Notes on the distribution of the frogs and toads of Kentucky. Amer. Midland Nat. 44 (3): 759-60.

———. 1950. The reptiles of Big Black Mountain, Harlan County, Kentucky. Copeia, No. 2, pp. 100-107.

———. 1953. The amphibians of Big Black Mountain, Harlan County, Kentucky. Copeia, No. 2, pp. 84-89.

———. 1955. The salamander *Eurycea longicauda pernix* in Kentucky. Copeia, No. 4, p. 309.

———. 1956. An additional record of the coal skink, *Eumeces anthracinus* (Baird), in Kentucky. Herpetologica 12 (3): 230.

———. 1956. The Salientia of Kentucky: Identification and distribution. Trans. Kentucky Acad. Sci. 17 (2): 81-86.

———. 1956. A study of the cottonmouth, *Ancistrodon piscivorus leucostoma,* in Kentucky. Trans. Kentucky Acad. Sci. 17 (1): 33-41.

———. 1957. A checklist and key to the amphibians and reptiles of Kentucky. Rev. ed. University of Kentucky, Lexington. 41 pp.

———. 1957. A distributional study of *Pseudacris nigrita* in Kentucky. Amer. Midland Nat. 57 (1): 129-32.

———. 1957. Observations on the mountain chorus frog, *Pseudacris brachyphona* (Cope), in Kentucky. Amer. Midland Nat. 57 (1): 125-28.

———. 1958. A deletion of distribution records of *Hyla squirella*. Copeia, No. 1, pp. 49-50.

———. 1960. A study of the worm snake, *Carphophis amoenus* Say, in Kentucky. Trans. Kentucky Acad. Sci. 21 (1-2): 10-16.

———. 1962. An aggregation of copperheads, *Agkistrodon contortrix*. Copeia, No. 3, p. 640.

———. 1963. On the activity of *Plethodon glutinosus* as influenced by light. Trans. Kentucky Acad. Sci. 24 (1-2): 13-19.

Barbour, Roger W., and Wm. L. Gault. 1952. Notes on the

spadefoot toad, *Scaphiopus holbrooki*. Copeia, No. 3, p. 192.

Barbour, Roger W., J. W. Hardin, J. P. Schafer, and M. J. Harvey. 1969. Home range, movements, and activity of dusky salamander, *Desmognathus fuscus*. Copeia, No. 2, pp. 293-97.

Barbour, Roger W., M. J. Harvey, and J. W. Hardin. 1969. Home range, movement, and activity of the eastern worm snake, *Carphophis amoenus amoenus*. Ecology 50 (3): 470-76.

Barbour, Roger W., and R. M. Hays. 1957. The genus *Desmognathus* in Kentucky; I. *Desmognathus fuscus welteri* Barbour. Amer. Midland Nat. 58 (2): 352-59.

Barbour, Roger W., and L. Y. Lancaster. 1946. Food habits of *Desmognathus fuscus* in Kentucky. Copeia, No. 1, pp. 48-49.

Barbour, Roger W., and E. P. Walters. 1941. Notes on the breeding habits of *Pseudacris brachyphona*. Copeia, No. 2, p. 116.

Barbour, Roger W., and W. A. Welter. 1941. Additions to the herpetofauna of north-eastern Kentucky. Copeia, No. 2, p. 136.

Barr, Thomas C., and R. M. Norton. 1965. Predation on cave bats by the pilot black snake. Jour. Mamm. 46 (4): 672.

Bishop, Sherman C. 1926. Records of some amphibians and reptiles from Kentucky. Copeia, No. 152, pp. 118-20.

————. 1941. Notes on salamanders with descriptions of several new forms. Occ. Papers Mus. Zool., University of Michigan 451: 1-27.

————. 1941. The salamanders of New York. New York State Mus. Bull. 324. 365 pp.

————. 1952. Handbook of salamanders. Comstock Publishing Company, Ithaca, New York. 555 pp.

Blair, W. F., A. P. Blair, P. Brodkorb, F. R. Cagle, and G. A. Moore. 1968. Vertebrates of the United States. 2nd ed. McGraw-Hill Book Co., New York. 616 pp.

Blanchard, Frank N. 1922. Amphibians and reptiles of western Tennessee. Occ. Papers Mus. Zool., University of Michigan 117: 1-18.

————. 1923. The snakes of the genus *Virginia*. Papers Michigan Acad. Sci., Arts, and Letters 3: 343-65.

————. 1925. A collection of amphibians and reptiles from southern Indiana and adjacent Kentucky. Papers Michigan Acad. Sci., Arts, and Letters 5: 367-88.

————. 1937. Eggs and natural nests of the eastern ring-neck snake, *Diadophis punctatus edwardsii.* Papers Michigan Acad. Sci., Arts, and Letters 12: 521-32.

————. 1938. Snakes of the genus *Tantilla* in the United States. Zool. Ser. Field Mus. Nat. Hist. 20 (28): 369-76.

————. 1939. A key to the snakes of the United States, Canada, and Lower California. Papers Michigan Acad. Sci., Arts, and Letters 4 (2): 65.

————. 1942. The ring-neck snakes, genus *Diadophis.* Chicago Acad. Sci. Bull. 7 (1): 5-144.

Blanchard, Frank N., and Frieda Cobb Blanchard. 1942. Mating of the garter snake *Thamnophis sirtalis sirtalis* (Linnaeus). Papers Michigan Acad. Sci., Arts, and Letters 27: 215-34.

Brandon, Ronald A. 1961. A comparison of the larvae of five northeastern species of *Ambystoma* (Amphibia, Caudata). Copeia, No. 4, pp. 377-83.

————. 1965. A second [record] of *Hemidactylium scutatum,* the four-toed salamander, in Kentucky. Trans. Illinois State Acad. Sci. 58 (2).

————. 1966. Systematics of the salamander genus *Gyrinophilus.* University of Illinois Press, Urbana. 86 pp.

Burkett, Ray D. 1966. Natural history of cottonmouth moccasin, *Agkistrodon piscivorous* (Reptilia). University of Kansas Publ., Mus. Nat. Hist. 17 (9): 435-91.

Burt, Charles E. 1931. A study of the Teiid lizards of the genus *Cnemidophorus* with special reference to their phylogenetic relationships. U.S. Nat. Mus. Bull. 154. 305 pp.

————. 1933. A contribution to the herpetology of Kentucky. Amer. Midland Nat. 14 (6): 669-79.

————. 1935. A key to the lizards of the United States and Canada. Trans. Kansas Acad. Sci. 38: 255-305.

Burt, Charles E., and M. D. Burt. 1929. A collection of amphibians and reptiles from the Mississippi Valley, with field observations. American Mus. Novitates 381: 1-14.

Bush, Francis M. 1957. A survey of the herpetofauna of Clemons Fork, Breathitt County, Kentucky. M.S. Thesis. University of Kentucky. 76 pp.

————. 1959. Food of some Kentucky herptiles. Herpetologica 15: 73-77.

————. 1959. The herpetofauna of Clemons Fork, Breathitt County, Kentucky. Trans. Kentucky Acad. Sci. 20 (1-2): 11-18.

Carpenter, Charles C. 1955. The garter snake. Scientific Monthly 81 (5): 248-52.

Carr, Archie. 1952. Handbook of turtles. Comstock Publishing Associates, Ithaca, New York. 542 pp.

———. 1963. The reptiles. Life Nature Library. Time, Inc., New York. 192 pp.

Chenoweth, W. L. 1949. Comments on some reptiles from the vicinity of Mammoth Cave National Park, Kentucky. Herpetologica 5: 21-22.

Clay, William M. 1938. A synopsis of the North American water snakes of the genus *Natrix*. Copeia, No. 4, pp. 174-82.

———. 1952. An albino snake (*Elaphe obsoleta*). Trans. Kentucky Acad. Sci. 13 (4): 285-86.

Clay, William M., Roberta Burckhardt Case, and Robert Cunningham. 1955. On the taxonomic status of the slimy salamander, *Plethodon glutinosus* (Green), in southeastern Kentucky. Trans. Kentucky Acad. Sci. 16 (3): 57-65.

Collins, Joseph T. 1962. Notes on *Virginia valeriae* Baird and Girard in Kentucky. Herpetologica 18 (2): 136-37.

———. 1962. *Ophisaurus attenuatus longicaudus* McConkey in Kentucky. Herpetologica 18 (3): 206.

———. 1964. Additional records of the scarlet snake, *Cemophora coccinea* (Blumenbach), in Kentucky. Jour. Ohio Herp. Soc. 4 (4): 106.

———. 1964. A preliminary review of the snakes of Kentucky. Jour. Ohio Herp. Soc. 4 (3): 69-77.

Collins, Joseph T., and C. J. Hirschfeld. 1964. *Lampropeltis doliata doliata* (Linnaeus) in Kentucky. Herpetologica 19 (4): 292-93.

Committee. 1956. Common names for North American amphibians and reptiles. Copeia, No. 3, pp. 172-85.

Conant, Roger. 1951. The reptiles of Ohio. Amer. Midland Nat. 20 (1): 1-284.

———. 1958. A field guide to reptiles and amphibians. Houghton-Mifflin Co., Boston. 366 pp.

———. 1963. Evidence for the specific status of the water snake *Natrix fasciata*. Amer. Mus. Novitates 2122: 1-38.

Cooper, John E. 1962. Cave records for the salamander *Plethodon richmondi richmondi* Pope with notes on additional cave-associated species. Herpetologica 17 (4): 251-55.

Cope, E. D. 1896. The geographical distribution of Batrachia and Reptilia in North America. Amer. Nat. 30: 886-902, 1003-26.

———. 1898. The crocodilians, lizards, and snakes of North America. U.S. Nat. Mus. Annual Rept. 1900: 153-1294.

Craddock, J. E., and W. L. Minckley. 1964. Amphibians and reptiles from Meade County, Kentucky. Amer. Midland Nat. 71 (2): 382-91.

Dunn, Emmett Reid. 1926. The salamanders of the family Plethodontidae. Smith College Fiftieth Anniversary Publications. 441 pp.

Dury, Ralph, and Wm. Gessing, Jr. 1940. Additions to the herpetofauna of Kentucky. Herpetologica 2 (2): 31-32.

———. 1940. Notes on *Lampropeltis* in Kentucky. Copeia, No. 4, p. 241.

Dury, Ralph, and R. S. Williams. 1933. Notes on some Kentucky amphibians and reptiles. Baker-Hunt Foundation Mus. Bull. 1: 1-22.

Eigenmann, C. H. 1899. Explorations of the caves of Missouri and Kentucky. Proc. Indiana Acad. Sci. 8: 59-61.

Endsley, J. R. 1954. An annotated listing of a herpetological collection mainly from Tennessee. Jour. Tennessee Acad. Sci. 29 (1): 36-41.

Ernst, Carl. 1965. Fang length comparisons of American *Agkistrodon*. Trans. Kentucky Acad. Sci. 26 (1-2): 12-18.

Ernst, Carl, and E. M. Ernst. 1969. Turtles of Kentucky. Jour. Int. Turtle and Tortoise Society 3 (5): 13-15.

Ernst, Carl, and B. G. Jett. 1969. An intergrade population of *Pseudemys scripta elegans* and *Pseudemys scripta troosti* in Kentucky. Jour. Herpetology 3 (1-2): 103.

Fitch, Henry S. 1954. Life history and ecology of the five-lined skink, *Eumeces fasciatus*. University of Kansas Publ., Mus. Nat. Hist. 8 (1): 1-156.

———. 1958. Natural history of the six-lined racerunner *(Cnemidophorus sexlineatus)*. University of Kansas Publ., Mus. Nat. Hist. 11 (2): 11-62.

———. 1960. Autecology of the copperhead. University of Kansas Publ., Mus. Nat. Hist. 13 (4): 85-288.

———. 1963. Natural history of the racer *Coluber constrictor*. University of Kansas Publ., Mus. Nat. Hist. 15 (8): 353-468.

———. 1965. An ecological study of the garter snake, *Thamnophis sirtalis*. University of Kansas Publ., Mus. Nat. Hist. 15 (10): 493-564.

Fitch, Henry S., and Harry W. Greene. 1965. Breeding cycle

in the ground skink, *Lygosoma laterale*. University of Kansas Publ., Mus. Nat. Hist. 15 (11): 565-75.

Funkhouser, Wm. D. 1925. Wildlife in Kentucky. Kentucky Geol. Surv., Frankfort, Kentucky. 385 pp.

———. 1945. Kentucky snakes. Dept. Ext., University of Kentucky. 31 pp.

Fuller, Tom C. 1959. Five new records of *Pituophis* in Kentucky. Herpetologica 15: 94.

Fuller, Tom C., and Roger W. Barbour. 1962. The distribution of *Cemophora coccinea* (Blumenbach) in Kentucky. Copeia, No. 3, pp. 637-38.

Garman, Harrison. 1894. A preliminary list of the vertebrate animals of Kentucky. Bull. Essex Inst. 26 (1): 1-63.

Gentry, Glenn. 1955. An annotated check list of the amphibians and reptiles of Tennessee. Jour. Tennessee Acad. Sci. 30 (2): 168-76.

———. 1956. An annotated checklist of the amphibians and reptiles of Tennessee. Jour. Tennessee Acad. Sci. 31 (3): 242-51.

Giovannoli, Leonard. 1936. *Scaphiopus holbrookii* in Kentucky. Copeia, No. 1, p. 69.

Gloyd, H. K. 1940. The rattlesnakes, genus *Sistrurus* and *Crotalus*. Chicago Acad. Sci. Spec. Publ. 4. 266 pp.

Goin, Coleman J., and Olive B. Goin. 1962. Introduction to herpetology. W. H. Freeman and Co., San Francisco. 341 pp.

Gosner, Kenneth L., and Irving H. Black. 1955. The effects of temperature and moisture on the reproductive cycle of *Scaphiopus holbrooki holbrooki*. Amer. Midland Nat. 54 (1): 192-203.

Grant, Chapman. 1936. Herpetological notes from Fort Knox, Kentucky. Proc. Indiana Acad. Sci. 45: 334.

Green, N. Bayard. 1938. The breeding habits of *Pseudacris brachyphona* (Cope) with a description of the eggs and tadpole. Copeia, No. 2, pp. 79-82.

———. 1941. The four-toed salamander in Kentucky. Copeia, No. 1, p. 53.

Green, N. Bayard, Paul Grant, Jr., and Bernard Dowler. 1967. *Eurycea lucifuga* in West Virginia: Its distribution, ecology, and life history. Proc. West Virginia Acad. Sci. 39: 297-304.

Hamilton, W. J., Jr. 1940. Observations on the reproduction behavior of the snapping turtle. Copeia, No. 2, pp. 124-26.

————. 1951. The food and feeding behavior of the garter snake in New York. Amer. Midland Nat. 46 (2): 385-90.

Hardin, S. W., J. P. Schafer, and R. W. Barbour. 1969. Observations on the activity of a seal salamander, *Desmognathus monticola*. Herpetologica 25 (2): 150-51.

Harris, Don R. 1965. A technique for collecting aquatic reptiles and amphibians. Jour. Ohio Herp. Soc. 5 (1): 35-36.

Hays, Rodney M. 1955. A comparison of *Desmognathus fuscus welteri* with several other members of the genus. M.S. Thesis. University of Kentucky. 61 pp.

Hecht, M. K. 1958. A synopsis of the mud puppies of eastern North America. Proc. Staten Island Inst. Art. Sci. 21: 5-38.

Hibbard, Claude W. 1936. The amphibians and reptiles of Mammoth Cave National Park proposed. Trans. Kansas Acad. Sci. 39: 277-81.

Highton, Richard. 1962. Revision of North American salamanders of the genus *Plethodon*. Florida State Mus. Bull. 6 (3): 235-367.

Highton, Richard, and Arnold B. Grobman. 1956. Two new salamanders of the genus *Plethodon* from the southeastern United States. Herpetologica 12: 185-88.

Highton, Richard, and Thomas Savage. 1961. Functions of the brooding behavior in the female red-backed salamander, *Plethodon cinereus*. Copeia, No. 1, pp. 15-98.

Hirschfeld, Corson J., and Joseph T. Collins. 1963. Range extensions for three amphibians in north-central Kentucky. Copeia, No. 2, pp. 438-39.

Huheey, James E., and Arthur Stupka. 1967. Amphibians and reptiles of Great Smoky Mountains National Park. University of Tennessee Press, Knoxville. 98 pp.

Hutchison, Victor H. 1956. Notes on the Plethodontid salamanders, *Eurycea lucifuga* (Rafinesque) and *Eurycea longicauda longicauda* (Green). Occ. Papers Nat. Speleo. Soc. (3): 3-24.

————. 1958. The distribution and ecology of the cave salamander *Eurycea lucifuga*. Ecol. Monog. 28: 1-20.

Klimstra, W. D. 1958. Some observations on snake activities and populations. Ecology 39 (2): 232-39.

————. 1959. Foods of the racer *Coluber constrictor* in southern Illinois. Copeia, No. 3, pp. 210-14.

Klimstra, W. D., and Frances Newsome. 1960. Some observations on the food coactions of the common box turtle, *Terrapene carolina carolina*. Ecology 41 (4): 639-47.

List, James C. 1950. Observations on the courtship behavior of *Thamnophis sirtalis sirtalis*. Herpetologica 6: 71-74.

Martof, Bernard S., and Robert L. Humphries. 1959. Geographic variation in the wood frog, *Rana sylvatica*. Amer. Midland Nat. 61 (2): 350-89.

McConkey, Edwin H. 1954. A systematic study of the North American lizards of the genus *Ophisaurus*. Amer. Midland Nat. 51 (1): 133-71.

McFarlan, Arthur C. 1950. Geology of Kentucky. University of Kentucky Press, Lexington. 531 pp.

McIntyre, T. 1926. Amphibia of Kentucky. M.S. Thesis. University of Kentucky. 86 pp.

Miller, Glenn M. 1964. A record length long-tailed salamander, *Eurycea longicauda longicauda* (Green). Jour. Ohio Herp. Soc. 4 (4): 106.

Minton, S. A., Jr. 1953. Variations in venom samples from copperheads (*Agkistrodon contortrix mokeson*) and timber rattlesnakes (*Crotalus horridus horridus*). Copeia, No. 4, pp. 212-15.

Mittleman, M. B. 1949. American Caudata. VI. The races of *Eurycea bislineata*. Proc. Bio. Soc. Washington 62: 89-96.

———. 1951. American Caudata. VII. Two new salamanders of the genus *Plethodon*. Herpetologica 7 (3): 105-12.

Neill, Wilfred T. 1957. The eggs of the crowned snake, *Tantilla coronata*. Herpetologica 13: 77-78.

Noble, G. K. 1931. Biology of the amphibia. McGraw-Hill Book Co., New York. 577 pp.

Parker, Malcolm V. 1939. Some amphibians and reptiles from Reelfoot Lake. Jour. Tennessee Acad. Sci. 12 (1): 60-86.

———. 1939. The amphibians and reptiles of Reelfoot Lake and vicinity, with a key for the separation of species and subspecies. Jour. Tennessee Acad. Sci. 14 (1): 72-101.

———. 1948. A contribution to the herpetology of western Tennessee. Jour. Tennessee Acad. Sci. 22 (1): 20-30.

———. 1951. Notes on the bird-voiced tree frog, *Hyla phaeocrypta*. Jour. Tennessee Acad. Sci. 26 (3): 208-13.

Peters, James A. 1964. Dictionary of Herpetology. Hafner Publishing Co., New York. 393 pp.

Pope, Clifford H. 1928. Some plethodontid salamanders from North Carolina and Kentucky with the description of a new race of *Leurognathus*. Amer. Mus. Novitates 306: 1-19.

———. 1946. Turtles of the United States and Canada. Alfred A. Knopf, New York. 343 pp.

———. 1947. Amphibians and reptiles of the Chicago area. Chicago Nat. Hist. Mus. 275 pp.

Rafinesque, C. S. 1832. Description of the *Spelerpes* or salamanders of the caves of Kentucky. Atlantic Jour. 1 (1): 22.

———. 1832. On the salamanders of the hills of east Kentucky. Atlantic Jour. 1 (1): 22.

———. 1832. On three new salamanders of Kentucky. Atlantic Jour. 1 (3): 121.

Reese, R. W. 1949. The occurrence of *Eumeces inexpectatus* in Kentucky. Nat. Hist. Misc. 39: 1-2.

Riemer, Wm. J. Ed. 1963. et. seq. Catalogue of American amphibians and reptiles. American Soc. Ichthyologists and Herpetologists, Bethesda, Md.

Risley, Paul L. 1933. Observations on the natural history of the common musk turtle, *Sternotherus odoratus* (Latreille). Pap. Michigan Acad. Sci., Arts, and Letters 17: 685-711.

Robinson, Thane, and K. Thomas Reichard. 1965. Notes on the breeding biology of the midland mud salamander, *Pseudotriton montanus diastictus*. Jour. Ohio Herp. Soc. 5 (1): 29.

Rossman, Douglas A. 1958. A new race of *Desmognathus fuscus* from the south-central United States. Herpetologica 14: 158-60.

———. 1962. *Thamnophis proximus* (Say), a valid species of garter snake. Copeia, No. 4, pp. 741-48.

———. 1963. The colubrid snake genus *Thamnophis*: A revision of the *sauritus* group. Florida State Mus. Bull. 7 (3): 99-178.

Savage, J. M. 1948. The pine snake in Kentucky. Copeia, No. 4, p. 302.

Schaaf, R. T., and C. Moore. 1969. Additional Kentucky records of *Hemidactylium scutatum*. Jour. Herpetology 3 (3-4): 180.

Schmidt, Karl P. 1953. Checklist of North American amphibians and reptiles. American Soc. Ichthyologists and Herpetologists, Bethesda, Md. 280 pp.

Schmidt, Karl P., and D. D. Davis. 1941. Field book of snakes. G. P. Putnam's Sons, New York. 365 pp.

Shoop, C. Robert. 1957. Eggs and young of the prairie king snake, *Lampropeltis c. calligaster*. Copeia, No. 1, pp. 48-49.

————. 1960. The breeding habits of the mole salamander, *Ambystoma talpoideum* (Holbrook), in southeastern Louisiana. Tulane Studies in Zoology 8 (3): 65-82.

————. 1965. Aspects of reproduction in Louisiana *Necturus* populations. Amer. Midland Nat. 74 (2): 357-67.

————. 1965. Orientation of *Ambystoma maculatum*: Movements to and from breeding ponds. Science 149 (3683): 558-59.

Smith, Albert G. 1948. Intergradation in worm snakes (*Carphophis*) from Kentucky. Nat. Hist. Misc. 18: 1-3.

Smith, Hobart M. 1946. Handbook of lizards. Comstock Publishing Company, Ithaca, New York. 557 pp.

————. 1950. Handbook of amphibians and reptiles of Kansas. University of Kansas, Mus. Hist. Misc. Publ. No. 2. 336 pp.

Smith, Philip W. 1961. The amphibians and reptiles of Illinois. Bull. Illinois Nat. Hist. Survey 28 (1): 298 pp.

Smith, P. W., and D. M. Smith. 1952. The relationship of the chorus frogs, *Pseudacris nigrita feriarum* and *Pseudacris n. triseriata*. Amer. Midland Nat. 48: 165-80.

Smith, P. W., and Hobart M. Smith. 1952. Geographic variation in the lizard *Eumeces anthracinus*. University of Kansas Sci. Bull. 34 (11): 679-94.

Snyder, David H., D. F. Burchfield, and R. W. Nall. 1967. First records of pigmy rattlesnake in Kentucky. Herpetologica 23 (3): 240-41.

Stein, Hilda A., and Philip W. Smith. 1959. The hellbender, *Cryptobranchus alleganiensis*, in Illinois. Copeia, No. 2, pp. 178-79.

Thurow, Gordon R. 1967. The aberrantly-pigmented grand canyon salamander, *Plethodon dorsalis*. Trans. Illinois State Acad. of Sci. 60 (3): 324-25.

————. 1968. On the small black *Plethodon* problem. Western Illinois University Series of the Biol. Sci. No. 6. 48 pp.

Tinkle, Donald W. 1957. Ecology, maturation, and reproduction of *Thamnophis sauritus proximus*. Ecology 38 (1): 69-77.

————. 1961. Geographic variation in reproduction, size, sex ratio, and maturity of *Sternothaerus odoratus* (Testudinata: Chelydridae). Ecology 42 (1): 68-76.

————. 1962. Variation in shell morphology of North American turtles. I. The carapacial seam arrangements. Tulane Studies in Zoology 9 (5): 331-49.

Trapido, Harold. 1944. The snakes of the genus *Storeria.* Amer. Midland Nat. 31 (1): 1-84.

Uzzell, Thomas M., Jr. 1964. Relations of the diploid and triploid species of the *Ambystoma jeffersonianum* complex (Amphibia, Caudata). Copeia, No. 2, pp. 257-300.

Walker, C. F., and W. H. Weller. 1932. The identity and status of *Pseudotriton duryi.* Copeia, No. 2, pp. 81-83.

Wallace, James T., and R. W. Barbour. 1957. Observations on the eggs and young of *Plethodon richmondi.* Copeia, No. 1, p. 48.

Webb, Robert G. 1962. North American recent soft-shelled turtles (Family Trionychidae). University of Kansas Publ., Mus. Nat. Hist. 13 (10): 429-611.

Weller, W. H. 1930. Notes on amphibians collected in Carter County, Kentucky. Proc. Junior Acad. Sci. (Cincinnati) 1 (5-6): 6-9.

————. 1931. Corrections to herpetological notices. Proc. Junior Soc. Nat. Sci. (Cincinnati) 2 (1): 7-9.

Welter, W. A., and Katherine Carr. 1939. Amphibians and reptiles of northeastern Kentucky. Copeia, No. 3, pp. 128-130.

Williams, John C. 1956. A study of *Plethodon glutinosus* (Green) in Kentucky. M.S. Thesis. University of Kentucky. 49 pp.

Williams, Kenneth L.,, and Larry D. Wilson. 1967. A review of the colubrid snake genus *Cemophora* Cope. Tulane Studies in Zoology 13 (4): 103-24.

Wright, A. H., and A. A. Wright. 1949. Handbook of frogs and toads. Comstock Publishing Company, Ithaca, New York. 640 pp.

————. 1952. A list of the snakes of the United States and Canada by states and provinces. Amer. Midland Nat. 48 (3): 574-603.

————. 1957. Handbook of snakes of the United States and Canada. Comstock Publishing Associates, Ithaca, New York. 2 vols.

INDEX